THE CLOCK OF THE NIGHT SKY

K.K. Birla Academy Monographs on the Cultural, Scientific and Technological Heritage of India (No. 2)

THE CLOCK OF THE NIGHT SKY

Dr. V. Krishnamurthy

UBSPD

UBS Publishers' Distributors Ltd.

New Delhi • Mumbai • Bangalore • Chennai •
Calcutta • Patna • Kanpur • London

UBS Publishers' Distributors Ltd.

5 Ansari Road, New Delhi-110 002
Phones : 3273601, 3266646 ☆ *Cable* : ALLBOOKS ☆ *Fax* : (91) 11-327-6593
e-mail: ubspd.del@smy.sprintrpg.ems.vsnl.net.in
Apeejay Chambers, 5 Wallace Street, Mumbai-400 001
Phones : 2046971, 2047700 ☆ *Cable* : UBSIPUB ☆ *Fax* : 2040827
10 First Main Road, Gandhi Nagar, Bangalore-560 009
Phones : 2263901, 2263902, 2253903 ☆ *Cable* : ALLBOOKS ☆ *Fax* : 2263904
6, Sivaganga Road, Nungambakkam, Chennai-600 034
Phone : 8276355 ☆ *Cable* : UBSIPUB ☆ *Fax* : 8270189
8/1-B, Chowringhee Lane, Calcutta-700 016
Phones : 2441821, 2442910, 2449473 ☆ *Cable* : UBSIPUBS ☆ *Fax* : 2450027
5 A, Rajendra Nagar, Patna-800 016
Phones : 672856, 673973, 656170 ☆ *Cable* : UBSPUB ☆ *Fax* : 656169
80, Noronha Road, Cantonment, Kanpur-208 004
Phones : 369124, 362665, 357488 ☆ *Fax* : 315122

First Published **1998**

Cover design : Ilaksha

Designed & Typeset at UBSPD in 11 pt Souvenir
Printed at Nutech Photolithographers, Delhi (India).

Dedicated
to
My Father
Sri R. Visvanatha Sastri
(1882-1956)
who implanted the nucleus of
the material of this book in my mind
when I was just a boy of eleven.

Foreword

The K.K. Birla Academy has been set up with the objective of presenting an image of India, one of the oldest civilizations in the world, by delineating its total and unified historical, developmental, scientific and cultural progress. A museum with visual exhibits, models, figures, diagrams and charts and using modern scientific audio-visual methods including robotics is in the process of being set up for the purpose. The exposition would highlight India's advance in science, technology, industry, arts, crafts and culture through different historical periods. This would be in comprehension of India's long and unbroken cultural traditions emanating from the distant past to the present and affording a peep into the future.

The Academy functions also as a focal point of research for disseminating authentic information about India's glorious past, its rich heritage and history and the scientific and cultural developments in the country. The Academy also sponsors research in these areas and encourages publication of research material.

The Academy has a scheme of publishing research monographs. The first monograph was "The Iron Pillar at Delhi" by Prof. T.R. Anantharaman.

This research monograph "*The Clock of the Night Sky*" is the second monograph published by the K.K. Birla Academy. It has been written by Prof. V. Krishnamurthy, who till recently was Director and Research Coordinator of the Academy. Prof. Krishnamurthy was formerly Deputy Director of the Birla Institute of Technology & Science, Pilani, and was also the President of Indian Mathematical Society. He is a well-known mathematician and a renowned scholar of Sanskrit and Vedic literature.

This monograph documents for the first time some of the rare Sanskrit formulae, 27 in number, which help to locate the position of the stars and make it possible to tell the time of the night by just a look at the star configuration near the Zenith in the sky.

The Monograph deals with the fascinating subject of stars and star gazing. I hope scholars in the field as well as scientists and all those interested in Indian heritage will find this monograph extremely useful.

8.3.97 **K.K. Birla**

Preface

It must have been some day in the latter half of July 1938. I was not yet eleven years old. I was walking along with my father and a few of his friends in what I thought was dead of night, to the river *Pennar* in *Cuddalore* in South India. We were all going for the daily morning bath in the river which was probably a mile away. I must have been complaining that I had been woken up too early in the morning. My father immediately told me: Look there in the skies right above your head. Don't you see the two stars which make up the two *Proshtapadas*? They are the two corners of a square, called the square of Pegasus. '*Ahirbudhne Mithune Bhanu*' declares my father with gusto. It looks like a magical incantation like ABRACADABRA to me. That means, says he, '*Ahirbudhna* is *Uttara-Proshtapada*. When that is above us, Mithuna is rising. And today the Sun is in the beginning of the *Karka* sign. That means it is already around 5-30 a.m. So why do you think it is dead of night?' I marvelled at his magic-like calculation. From that date onwards, for a whole year, as we walked along for our morning walk to the river, he taught me how to recognise the 27 asterisms and how by recognising them almost directly above us in the sky we can tell the time of night, without the help of any clock or wrist-watch. In a few days I learnt from him the whole set of 27 formulae that he seemed to remember as his second nature. I also learnt about their use. But it took me a long time after that, well beyond even my college days of Astronomy, before I could understand the 'why' of these formulae. This book is an attempt to share with the modern reader the thrill of this understanding.

It was Dr. S. Venkateswaran, Director, BITS, Pilani, who, some years ago prompted me to write this book. However it was only last year that I took his suggestion seriously. But once I started working on the Project, it opened out to me several fascinating avenues which I had not anticipated. The work proved to be the right choice for research under the Academy whose main focus is the heritage of India, particularly in Science and Technology.

The book has actually turned out to be an easy introduction to, and at the same time an interesting adventure into, the mysteries of the star-studded night sky. The first chapter is just an introduction, useful to (and necessary for) a total stranger to astronomy and its language. The second chapter introduces the 27 stars of the Indian tradition. Modern usage however associates these stars exclusively to astrology. This is a quirk of history. This book is a book on astronomy and not on astrology. And one of the purposes of the book is to focus attention on these 27 stars totally from an astronomical point of view. The third, fourth and fifth chapters are the main content of what the book started to unravel. It is in these chapters that we give the various formulae that provide the clock of the night sky. Four different sets of formulae are discussed — two from the Sanskrit language and two from Tamil. Except for one of the two sets of Sanskrit formulae — which was actually a later discovery in the form of an eighteenth century Tanjore manuscript — the other three do not seem to have any written tradition, as far as this author knows. A number of examples illustrate the application of the formulae. The last chapter subjects all these to a critical analysis.

It is hoped that the book will be read not only by the young, by the layman and by novices but also by the old, by the scholars and by the experts. The author will feel highly rewarded if the book provokes a wider and deeper interest in Astronomy in general and the mysteries of the stars in particular.

It is a pleasure to place on record my deep-felt gratitude to Shri K.K. Birlaji who thought it fit to call me to the Academy and support me with all the physical and other facilities so that this book could be written.

1 Nov. 1997

V. Krishnamurthy
4, Papani Apts.,
6, 67th Street,
Ashoknagar, Chennai-600 083

Acknowledgements

Acknowledgements with gratitude are due to

1. Dr. S.S. Janaki, Advisor, Kuppuswami Sastri Research Institute, Madras, and
2. Dr. J.V. Narlikar, Director, Inter University Centre for Astronomy, Pune for providing me with the right names to contact;
3. Prof. L. Narayana Sharma, Professor of Astrology, Samskrit College, Madras for locating the Tanjore manuscript for me;
4. Dr. Kapila Vatsyayan, Academic Director, IGNCA, New Delhi for providing me with the right names to contact and for making available from the Library of the IGNCA a copy of the original version of the mss. referred to in No.3;
5. Dr. T.A.V. Murthy, Librarian, Kalanidhi, IGNCA, New Delhi and
6. Dr. Sampat Narayan , IGNCA New Delhi for help in reading the Tanjore mss.;
7. Professor R.K. Kochar, Indian Institute of Astrophysics, Bangalore for expressing great interest and for identifying some references;
8. Dr. Mrs. Nirupama Raghavan, Director, Nehru Planetarium, New Delhi for encouragement throughout the writing, for allowing the use of models from the Planetarium and for arranging a lecture-cum-demonstration on June 29, '96 at the Nehru Planetarium which served as a curtain-raiser for the book ;
9. Dr. A.K. Bag, Indian National Science Academy, New Delhi
10. Dr. B.G. Siddharth, B.M. Birla Science Centre, Hyderabad,
11. Ra Ki Rangarajan, Writer, Madras and
12. Makaram Kalyanaraman, Madras for expressing keen interest in the book;
13. Prof. K.E. Raman & Ms. Usha V. Subramanian, BITS Pilani for helping me on the computerised charts of Appendix II;
14. Staff of the Academy for all infra-structural help, and in particular, Miss S. Radha, Project Assistant at the Academy, who worked with me on all the examples, worked meticulously on all the figures of the book on the computer and in general proved to be my testing ground; and finally, most of all, to Shri S.M. Agarwal, Member, Governing Body of the Academy, for giving me the opportunity to work in the Academy and for supporting me with all the facilities.

CONTENTS

Chapter 1

Introductory

There are many pursuits which are mind-elevating to Man—listening to music, thinking of the divine, selfless service to other living beings, inquiring about man's innermost validity, enjoying poetry and the like. But it can be safely asserted that none of these was his first occupation in the historical evolution of *homo sapiens* as a civilized being. Believe it or not, the first 'human' activity which occupied his attention must have been 'star gazing'. Any other activity, like looking for a shelter for himself, finding his food, mating and procreating, were all nothing but a continuation of the same animal instinct with which he lived for ages in his earlier evolutionary stages. The first thing he did *as Man*, was to look at the Sun every day and at the Moon and the stars every night, in awe and wonder, observe, ... and, THINK! And in that process, he must have asked himself the questions, What? Why? and How? In this first scientific exploration of man, he, it appears, arrived at the concept of time and measurement of time by linking it with the regularity of the celestial procession of the objects that he saw in the sky. It was thus that the science of Astronomy was born. Every school kid knows about the fundamentals of astronomy by means of his visits to Planetaria and his study about the planets and the Sun at school. But what is usually not known or not taught to him is the unique contribution of the ancient culture of India to this science. In this short monograph we shall deal with one small fragment of this contribution—that fragment which is not even recorded in any of the modern books. We call this **'The Clock of the Night Sky'.**

The ancients of India have passed on to us 27 mathematical formulae coded in the Sanskrit language, but not very difficult to remember. In fact, very possibly, it has mostly come down to us by word of mouth and oral transmission from generation to generation as it came to this author from his father. These formulae enable one to tell the time of night by just one look at the stars right above us. What is amazing here is the fact that as early as the start of human civilisation man could not only recognize the patterns of the different asterisms and some constellations, but could also keep track of them night by night so as to register in his mind the changing patterns of the sky map and further, formulate these in the form of short aphorisms which would tell the time of night in relation to what one sees above oneself in the sky. This amazing feat of Early Man is the content of this book. However we shall not take the arduous path of duplicating or reconstructing the same path as he did in discovering these formulae. Obviously it must have taken him several years, probably a few generations, before he arrived at these formulae. In the light of the knowledge about the stars and planets that we possess today, we shall only try to interpret the formulae that early man enunciated and we shall see how accurate and scientific he has been in evolving these formulae. In the end we shall also see how there are corresponding formulae in the Tamil world, which seem to have had an independent source. They actually attempt to give more information about the stars *per se*. In this introductory chapter, however, we shall describe some elementary facts of Astronomy so that we may be able not only to translate the Samskrit formulae in the scientific language which is commonly understood but also apply them properly. So here we go, into basics of (High School) Astronomy!

Though the planets, the stars and the Sun are all at different distances from the Earth, it is convenient to imagine that all of them lie on an imaginary sphere (called the ***celestial sphere***) of very large radius with the Earth at the centre. However, different observers on the Earth will see different parts of the celestial sphere, because of the location from which they are seeing. Also the same observer at a single location will see different views of the celestial sphere at different times of the night. Since this book is being written for the understanding of even the layman, we shall not assume any knowledge on his part of the technical jargon of astronomy. So we shall introduce almost all the terms that we might use.

Once we can picture the sky as a globe, we can fix positions on it by the same means that we describe positions on earth. If we see the globe map of the Earth carefully, we see two different types of circles drawn on its surface. Those that pass through both the North and South Poles are called ***meridians***. The other circles cut the meridians at right angles and go around the Earth parallel to each other. These are called ***parallels of latitude*** and the one that is halfway between the Poles is called the ***equator.***

Any circle (on the sphere) that has as its centre the centre of the sphere is called a ***great circle***. All the meridians and the equator are great circles on the Earth. The parallels of latitude other than the equator are not great circles; they are called ***small circles***. Note that great circles of a sphere can also be described as sections of the sphere obtained by cutting the sphere by a plane passing through the centre of the sphere. See Fig.1. In the same manner small circles are sections of the sphere by planes which do not pass through the centre of the sphere.

Now let us look at the celestial sphere. See Fig 2. First there is *the* ***celestial equator***. It is the great circle *QR* which is the section of the celestial sphere by the plane of the terrestrial equator. The terrestrial equator *qr* on the Earth is itself a great circle on the Earth. When we extend it in all directions beyond the Earth and imagine it as cutting the celestial sphere, what we obtain is the celestial equator *QR*. It need not be emphasized that these are imaginary lines concocted only for the purpose of convenience of communication.

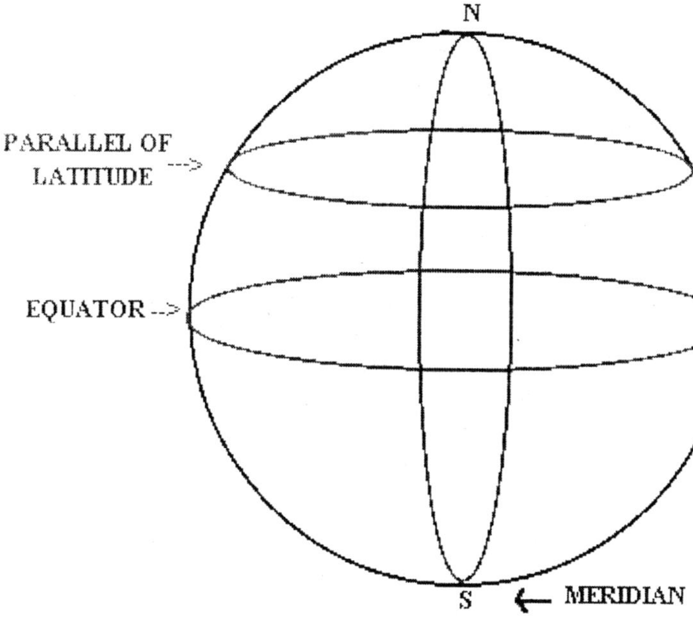

FIG 1

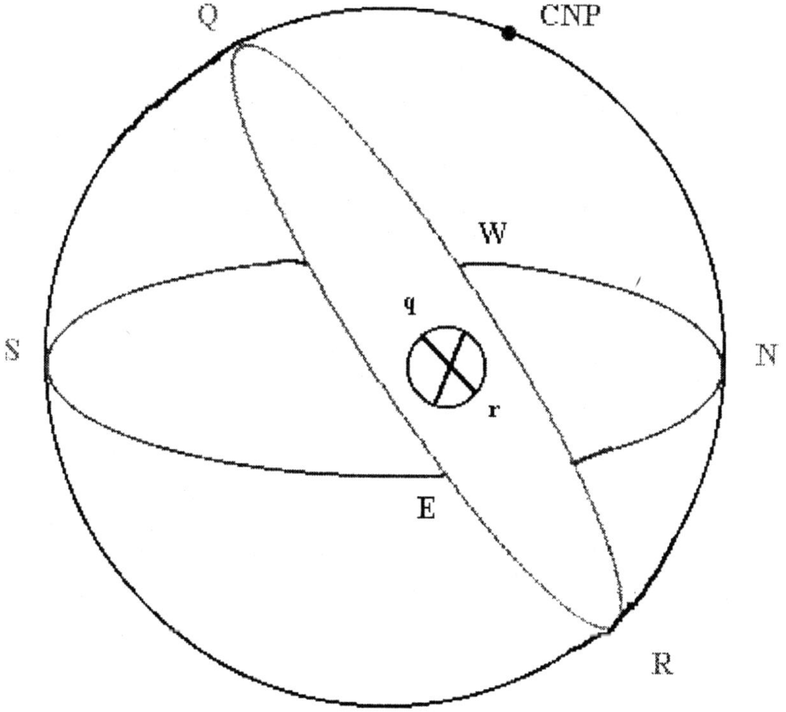

CELESTIAL SPHERE

FIG 2

If our view is unobstructed, we see the Earth as a flat circular disc. The border of this disc is the *horizon*. In the centre of the disc is ourselves—the *observer*. Wherever we may be we are always at the centre of this disc which is visible to us. The distance of a star from the horizon, is called its ***altitude***. Remember, distances on the celestial sphere — or for that matter any sphere — are measured in terms of the measure of the angle subtended at the centre. Let us denote by *A* the position of the observer on Earth. The point on the sky which is directly above A is called the Zenith for the observer *A*.

The axis *pp'* on the Earth which joins the North Pole (*p*') and the South Pole (*p'*)is the axis around which the Earth rotates daily once every 24 hours. Extend this axis in both directions to meet the celestial sphere in the imaginary points *P* and *P'*. These are called the Poles of the celestial sphere. Fortunately there is a star almost right at the position of the celestial North Pole. This star is therefore called the ***Pole Star***. For any observer in the northern hemisphere of the Earth the Pole Star is due north. Its altitude above the horizon however will depend on how far north of the equator (on Earth) the observer is. Precisely, if the observer A is Ø degrees north of the equator (that is, his latitude is Ø degrees north) the altitude of the Pole Star above the horizon is also Ø degrees. The why of this is explained in Appendix I. Once we have recognised this observation, we may hereafter, for the purpose of this book, take the point of observation to be the centre of the celestial sphere, thus ignoring the Earth's radius. The great circle passing through the Zenith *Z* and the celestial Pole *P* is called the ***celestial meridian***, or simply, ***meridian***, if no confusion can arise. It passes through the North and South points of the horizon. See Fig. 3.

The great circle passing through *Z* and *Z'* (the point opposite *Z* at the other end of the diameter through Z) and at right angles to the meridian is called the ***prime vertical*** and it meets the horizon at the East and West points (*E* and *W*). The entire celestial sphere rotates once every day around the axis *PP'* at the rate of 15° per hour in a direction opposite to that of the Earth's diurnal rotation about itself. Here the number 15 is because the Earth rotates through 360° in a period of 24 hours. Because of this rotation every star in the sky appears to rotate around *P* through a full circle in a full day and night. The sky is therefore changing, slowly but continuously, as the stars revolve round the Pole. Stars which we saw an hour ago in the west have set now. Stars which we saw almost above us an hour ago now are lower down in the west. Stars which we saw an hour ago in the east have risen further and there are new stars in the east. This (east-to-west) procession of the stars goes on and on without end. So we see a different sky every hour.

Not only do we see changes every hour, but if we carefully observe the night sky from night to night, we will see a change from month to month and from season to season. The sky in the month of March will look different from the sky in, say, June, or September though the time of observation is the same. This is because the revolution of the stars around the Pole does not take a full 24 hours but only 23 hours and 56 minutes. This shortfall of 4 minutes is what makes the sky different from night to night and month to month. Four minutes time is not much when isolated from the context. But slowly it adds up.The change from night to night may not be perceptible to the naked eye, but certainly

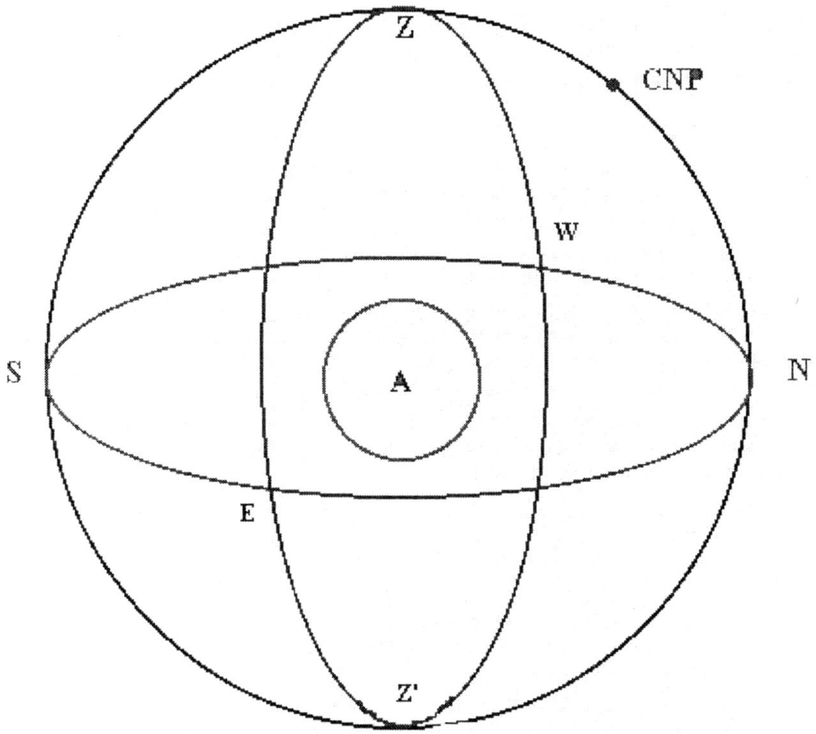

<-- PRIME VERTICAL
CELESTIAL MERIDIAN
FIG 3

it becomes very perceptible even to the casual observer from month to month, and definitely, from season to season. Since 30 x 4 minutes is two full hours, thirty days from now, we will see the same sky *two hours earlier*! Or in other words, one month from now the same stars will rise two hours earlier. A star is at its highest point when it crosses the meridian (from east to west). This highest point is called its ***upper transit***. Suppose a star transits the meridian tonight at 11 p.m.. One month ago the same star would have crossed the meridian at 1 a.m. One month from now the same star will cross the meridian at 9 p.m. Two months from now it will cross the meridian at 7 p.m. Six months from now it will cross the meridian at 11 a.m. and therefore its upper transit would not be visible. Twelve months from now it will cross the meridian again at 11 p.m. Thus the sky maps of the stars keep changing continuously within each night and from night to night but will revert to the same position in exactly a year from now.

But halt! All this movement of the stars is only apparent. The stars are not moving from night to night or month to month. It is because of the diurnal motion of the Earth around the sun that the stars seem to change their positions during a single night and it is because of the annual motion of the Earth around the Sun that the stars seem to change their positions from night to night and month to month. Now the Sun itself is a star. So as a star it has the same apparent change of position from month to month. The Sun happens to be the nearest star. The next nearest star is four light years away. The Sun is so near to us compared to the distant stars that the change in the position of the Sun from month to month is very pronounced *in the background of the distant stars*. If we keep track of the positions of the Sun throughout the year, in the backdrop of all the other stars which are millions of miles away, we will be able to mark off on the celestial sphere a great circle, called the ***ecliptic***. The ancients have been able to recognise this path of the Sun in the course of each year and they have divided the regions near the ecliptic into twelve constellations, called the twelve *Râsis* in all Indian books. These twelve constellations together called the *Zodiac* have Sanskrit names. Their corresponding English names which go back to the Greek civilisation are given in the Table below in parentheses against each.

Mesha (**Aries**)	*Tulâ* (**Libra**)
Vrishabha (**Taurus**)	*Vriščhika* (**Scorpio**)
Mithuna (**Gemini**)	*Dhanus* (**Sagittarius**)
Karka (**Cancer**)	*Makara* (**Capricornus**)
Simha (**Leo**)	*Kumbha* (**Aquarius**)
Kanyâ (**Virgo**)	*Mîna* (**Pisces**)

The Sun takes about a month to travel through each of these constellations. The popular way of saying this is that the Sun 'stays' for a month in each of these constellations. The path of the Sun through these constellations of the Zodiac is the ecliptic, which is only an imaginary great circle on the celestial sphere, just as the celestial equator is. This line across the sky is not only the annual path of the Sun but *more or less* also the path of all the planets in their revolution around the Sun. And further, since the Moon orbits the Earth at only a slight angle to the plane of the orbits of the planets, it too always appears close to the ecliptic. The two great circles, the ecliptic and the equator, meet at two points, which have standard notations for them, viz., γ the First Point of Aries and Ω the First Point of Libra respectively. These two points, called the **equinoxes**, are fixed on the ecliptic and the equator. The point where the Sun crosses the equator (in its annual motion) from south to north is called the **vernal equinox**. This happens on March 21. The other one (where the Sun crosses the equator from south to north) which happens on September 23 is called the **autumnal equinox**.

To identify a celestial object, particularly a star, astronomers use this reference point γ on the equator. The distance α of the object along the equator, measured from γ in the anticlockwise direction is called its **right ascension**. The distance δ (north or south) of the object from the equator is called its **declination**. See Fig.4. This means that if the *celestial* longitude circles are drawn from the Celestial North Pole (CNP) and if the celestial longitude circle passing through the celestial object S meets the equator at M then $\gamma M = \alpha$ and $SM = \delta$. Similarly if ENP is the ecliptic North Pole, that is the point 90 degrees north of the ecliptic and parallels of longitude circles are drawn through ENP (which are all perpendicular to the ecliptic (see Fig 5), and the longitude circle going through the celestial object S meets the ecliptic at F then $\gamma F = \gamma$ and $SF = \beta$. γ and β are respectively called the *longitude* and *latitude* of the celestial object S.

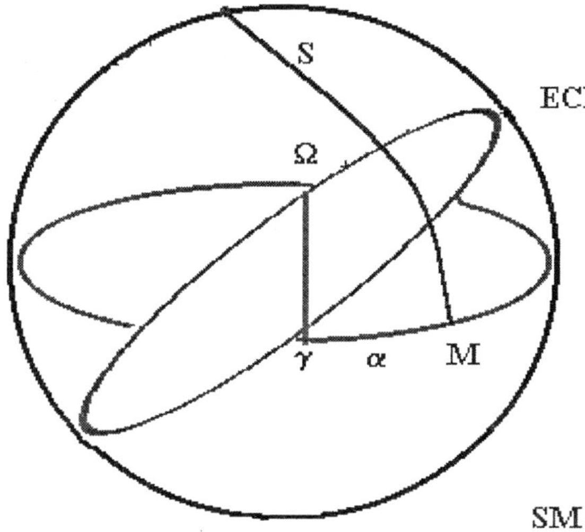

FIG 4

Thus the pairs (α, δ) and (λ, β) are two pairs of permanent coordinates of a star just as (terrestrial) longitude and latitude are permanent coordinates of places on Earth. In particular, the right ascension and declination of a star do not change over the course of the year nor do they change from place to place of observation on Earth. On the other hand the right ascension and declination of the planets, the Sun and Moon change through the days of the year. Here α ranges from 0 degrees to 360 degrees and δ ranges from 0° to 90° North or 90° South. We can also express the range of the right ascension α in terms of hours, from 0 hours to 24 hours. Since the sky seems to turn overhead once every 24 hours, we measure right ascension in hours. The 360° turn of the celestial equator once every 24 hours is divided therefore into 24 hours of right ascension. This will give it 1 *hour of right ascension* for every 15 degrees of turn. Consequently each *minute of right ascension* is 1/60 of 15° and this equals 15 minutes of arc. Again, each *second of right ascension* is 1/60 of 15′ and this equals 15 seconds of arc.

Recall that the celestial sphere is not a real sphere or a globe in the sense that the Earth is. Strictly speaking there is no such thing as the sky — only stars on a background

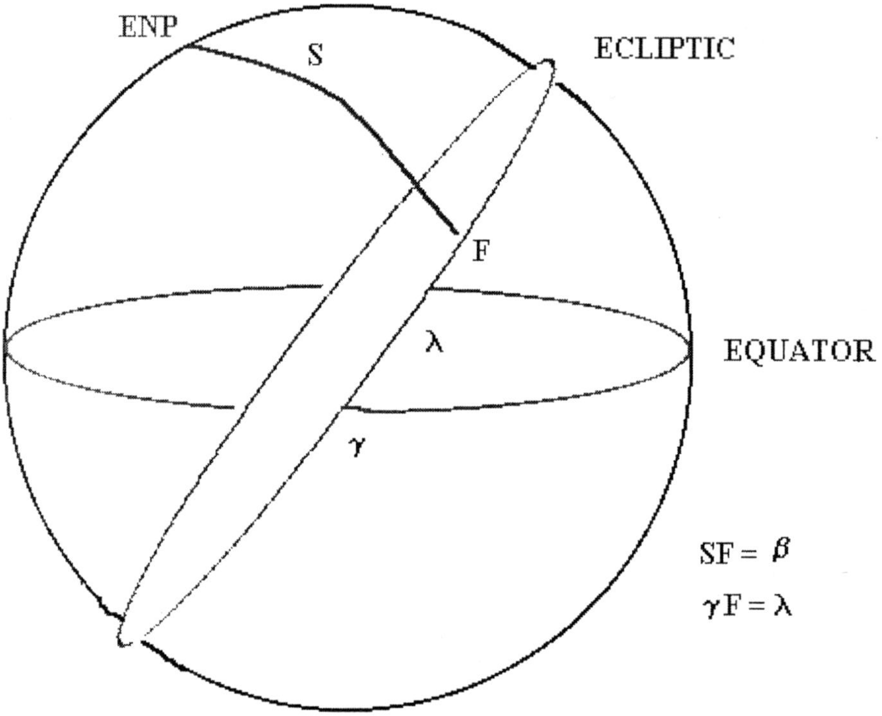

FIG 5

which may be called a void. We actually see one half of the hollow globe at any moment, with the other half slowly coming to view as the celestial sphere rotates. As the celestial sphere slowly rotates from East to West, the observer inside sees the star move upward at an angle. At its highest point, it is exactly on the meridian. After crossing the meridian, it descends slowly on the western half of the sky. It goes down with the same slant as when it rose, only in the opposite sense and sets in the western horizon. And as the celestial sphere keeps on turning, the star travels below the horizon; the observer cannot see it until it rises again in the eastern horizon.

When the Sun crosses the meridian, we have noon. The average time span from one noon to the next noon is one day, the ordinary day, which for us means 24 hours. The astronomer calls this the solar day. He has also the sidereal day which is the average time span from one sidereal noon (the time when a fixed star crosses the meridian) to the next sidereal noon. From one upper transit (called culmination) of a given star to the next upper transit of the same star, the star apparently has made a full turn of the celestial sphere. Actually it is the Earth that does the turning. The sidereal day is therefore very nearly the true period of rotation of the Earth with respect to the stars. This sidereal day is four minutes shorter than the solar day. The interval between two successive upper transits of the Sun is the Solar Day. It is longer than the sidereal day by about 3^m 56^s. This is because as the Earth turns once around its axis, it returns to the same point with respect to the stars, whereas in the meantime by its own annual motion around the sun, it has allowed the apparent Sun to move on the celestial sphere a distance of $3^m\ 56^s$ (= 4^m approximately), with which it has to catch up, in order to align itself with the Sun in the same position as the previous day. Thus sidereal day is approximately $23^h\ 56^m$ and solar day is 24^h.

There are standard tables available for conversion between sidereal time and solar time for each day of the year. The sidereal time of day is calculated from ***sidereal midnight*** which is the time when γ has its ***upper transit***. So the sidereal time at any instant is the number of hours since γ has crossed our meridian. There are two ways of determining this. One is to look up the table of conversions mentioned above.

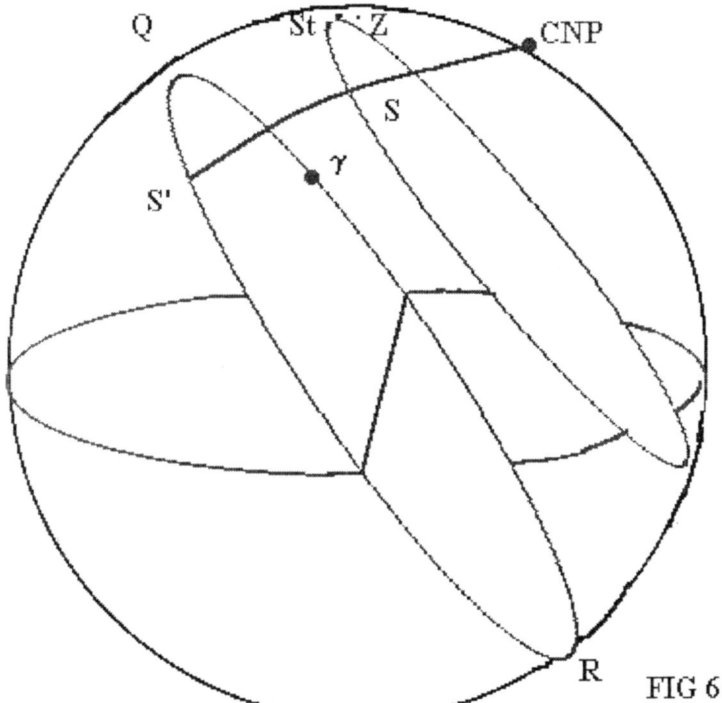

FIG 6

Alternatively, from Fig.6 we see that $\gamma QS' = \gamma Q + QS'$. This means: **Right Ascension of Star S = Sidereal time at the instant + hour angle of the star at the instant.** If the star happens to be in position St i.e., its upper transit, then the sidereal time at that instant and the right ascension of the star will coincide. Thus we have come to an important fact of astronomy: *The right ascension of a star is equal to the sidereal time of its upper transit*. Since the right ascension of every star is easily available in any star catalogue, the sidereal time can be calculated. Conversely if you know your sidereal time you know which stars are most favourably placed for observing. They are those which are within a few hours of the current sidereal time, provided they are in the right range of declination. This is the observation which the Hindu ancients seem to have exploited. The 27 formulae tell you the following: When such and such a star is on the meridian or near the meridian, what constellation of the Zodiac is rising in the East? Since the Sun's position in the Zodiac on any day of the year is known, the distance in hours between the constellation that is rising and the constellation in which the Sun is can be calculated mentally. This is the key point which makes the formulae work for finding the time of night at any instant by a look at the stars. The formulae are 27 in number because the ancient Hindus have identified 27 asterisms which lie more or less in the regions of the zodiac. For each such asterism, there is one formula that tells which zodiacal sign is rising when that asterism is in or near the meridian. So our next task now is to be able to identify these 27 asterisms. This shall be the business of the next chapter. .

Chapter 2

The 27 Nakshatras of the Hindu Calendar

Before we get further into the book it is necessary to forewarn the reader that ***this book is not about astrology in any sense of that word***. The book is on astronomy, pure and simple, however much we may use Indian names of stars and signs of the Zodiac, which names are usually heard by the layman only in connection with astrology of the Indian variety.

We know that each of the twelve constellations of the Zodiac spans about 30° or two hours of the longitude range on the ecliptic. The Hindu ancients divided each of these constellations into two and a quarter units of a Nakshatra-span, thus giving rise to 27 Nakshatras. Derived from *nakta-tra*, the word nakshatra means 'the guardian of the night'. Each Nakshatra is either a single star or a group of stars. We shall refer to each Nakshatra as an asterism. These asterisms are given below in the order of succession in which the signs of the Zodiac are listed.

Ašvinî	*Maghâ*	*Mûla*
Bharanî	*Pûrvaphalguni*	*Pûrvashâdhâ*
Krittikâ	*Uttaraphalguni*	*Uttarashâdhâ*
Rohinî	*Hastâ*	*Šravana*
Mrigaširah	*Chitrâ*	*Šravishthâ*
Ârdrâ	*Svâti*	*Šatabhishaj*
Punarvasu	*Višâkhâ*	*Pûrva-proshtapadâ*
Pushya	*Anurâdhâ*	*Uttara-proshtapadâ*
Âšleshâ	*Jyeshthâ*	*Revatî*

These are the Sanskrit names of the asterisms. A good familiarity with these 27 asterisms in terms of their locations and formations in the map of the sky is necessary for applying the formulae. It is unfortunate that in India we talk so highly about our heritage but in many practical situations we tend to do things which only make us forget this heritage. The fact that these 27 nakshatras have been known and used in all secular, religious, academic and ritualistic situations is ignored by many of us in public. For example, the media which talk of, write about, or show maps of the night sky do not at all refer to the standard Indian names of the Nakshatras. It is all carried on as if everything has been learnt only from the western scholars. What is more, even the mythology associated with individual stars is only taken from western books, as if no mythology existed in this part of the world. And the irony of it all is that much that was known in India was known long before most of the other part of the world had any civilisation of their own.

For a beginner of star reading in the northern hemisphere the first thing to learn to locate is the ***Sapta-Rishi*** constellation, known in English as the **Big Dipper** which is itself in the major constellation called **Ursa Major**. The *Sapta-Rishi* is made up of seven stars, which are known in Indian mythology as the Seven Great Sages of Hindu Mythology. These are[1] *Atri, Vašishta, Kašyapa, Gautama, Bharadwâja, Višwamitra* and *Jamadagni*. The Big Dipper is in the form of a bowl (rectangular) made up of four stars and three stars in the curved handle. The two stars in the bowl away from the handle are called the **Pointers**. They point to Polaris which is almost at the Celestial Pole. The middle star in the handle is ***Vašishta*** called *Mizar* in Western terminology. Alongside Mizar there is a tiny faintly shining star known by its Arabic name **Alcor**. In India it is more well-known as ***Arundhati***, the model of chastity in Hindu mythology and the wife of *Vašishta*. In all Hindu marriages one of the rituals is for the bride to see this star *Arundhati*. Whether she sees it or not, she is asked to reply, 'I see' when she is shown the star Vašishta and the question is put to her: 'Do you see Arundhati closeby?' — the whole performance being suggestive of a vow of, and prayer for, firmness in conjugal life. *Mizar* and *Alcor* together form the most famous and readily observable double star. The angular distance between the two stars is close to 12 minutes of arc, which is slightly more than a third of the apparent lunar disc. However the apparent proximity of these two stars is due only to their unimaginable distances from us. Actually they are separated from each other by a distance greater than 17,000 times that of the Earth from the Sun — about 2.5×10^{12} kilometres. But on the cosmic scale this is no distance at all. In fact *Arundhati* is quite close to *Vašishta*, 16 times closer than the Sun is to Alpha Centauri, our nearest star, (after the Sun) which is four light-years away. Except for this vague familiarity with the star *Arundhati* and *Vašishta*, the lay Indian may not recognise the 27 asterisms of Indian astronomy and astrology in the physical sky above him. So we shall describe each one of these in a little detail.

First let us have a few words about the scientific concept of magnitudes of brightness of stars. There is a description of scale of magnitudes coming down to us from the second century B.C. The Greek Astronomer Hipparchus called the brightest stars, ***stars of the first magnitude***. The next brightest stars were called by him ***stars of the second magnitude*** and so on. The faintest stars visible to the naked eye were called by him

1. This is according to the Vishnu Purana.

stars of the sixth magnitude. This scale was placed on a mathematical basis in the nineteenth century. A difference of 5 magnitudes corresponds to a factor of 100 in brightness. A few stars are even brighter than first magnitude and they have been accommodated by having magnitudes of zero and also negative numbers. The brightest star in the sky is Sirius (*Sanskrit* name used by Hindus: *Mrigavyâdha*) Its magnitude is –1.4. The second brightest star Canopus (Sanskrit name: *Agastya*) visible in the southern skies, is of magnitude –0.7. For each magnitude added, stars are about 2.5 times fainter in brightness. For each magnitude subtracted, stars are 2.5 times brighter.The smaller the magnitude number, the brighter is the object. For instance, the full moon has magnitude –12.6 and the Sun itself -26.8. The stars within the same constellation are usually indicated by the Greek letters in their alphabetical order, with α standing for the brightest star in the constellation, β for the next brightest and so on. In describing the various asterisms all the remarks about best time for the visibility of the asterism apply to northern latitudes on Earth. Also note that as far as the same night is concerned, the later you see a star the more 'west' it is. As far as different days are concerned, on the later day (same time), the star is more west. If you want to see it at the same place one month later, you have to observe it two hours earlier. The following Table would be of help.

Thumb rule for when and where to look for same star.

Observed Star was/will be	Date of observation	Time of observation
In the Same Position	One month later	Two hours earlier
In the Same Position	One month earlier	Two hours later
Further West by 30°	One month later	**Same time**
Further East by 30°	One month earlier	**Same time**
Further West	**Same day**	Later
Further East	**Same day**	Earlier

Now comes the list and description of the 27 Nakshatras.

Ašvinî: This is the brightest star in the constellation Aries. Its technical name is α-Arietis. It is also popularly known as **Hamal**. It can be spotted near the meridian around 22 hours on December 1, around 04 hours on September l. and accordingly at other times and nights as per the thumb rule in the foregoing table.

Bharanî: This is a faint star in Aries. In the hierarchy of magnitude within Aries, it is τ–Arietis. If conditions are favourable it may be spotted near the meridian around 22 hours on or about December 10 and around 04 hours three months earlier.

Krittikâ: This is a cluster of six stars well known as Pleides. There is an interesting Hindu mythological story about this. It appears the six wives of six of the seven universal Rishis of this age received the Energy of Lord Šiva, through a miraculous happening involving the involuntary intervention of the Lord of Fire. They conceived and brought forth six children all of whom combined into one child of six faces. This is the deity *Shanmukha*. The six husbands cursed their wives to become a cluster of stars and that is the Pleides cluster of six *Krittikâ* sisters. Because they nursed *Shanmukha* as a child, the latter is also known as ***Kârtikeya***. Western mythology talks of this cluster as the Seven Sisters of Greek Mythology. The cluster is actually made up of a large number of stars, the number going up to hundred with more and more powerful telescopes.

The Pleides cluster can be seen around December 20 near the meridian at 2200 hours and at 04 hours three months earlier. See Fig 1 for its location[2] with respect to neighbouring stars. Note that there is a fundamental difference between star maps and earthly geographical maps. In the latter the north direction is to the left of East, if you face east. In star maps it is the reverse. Because star maps are supposed to be held aloft over your head with the north pointing correctly to the north.

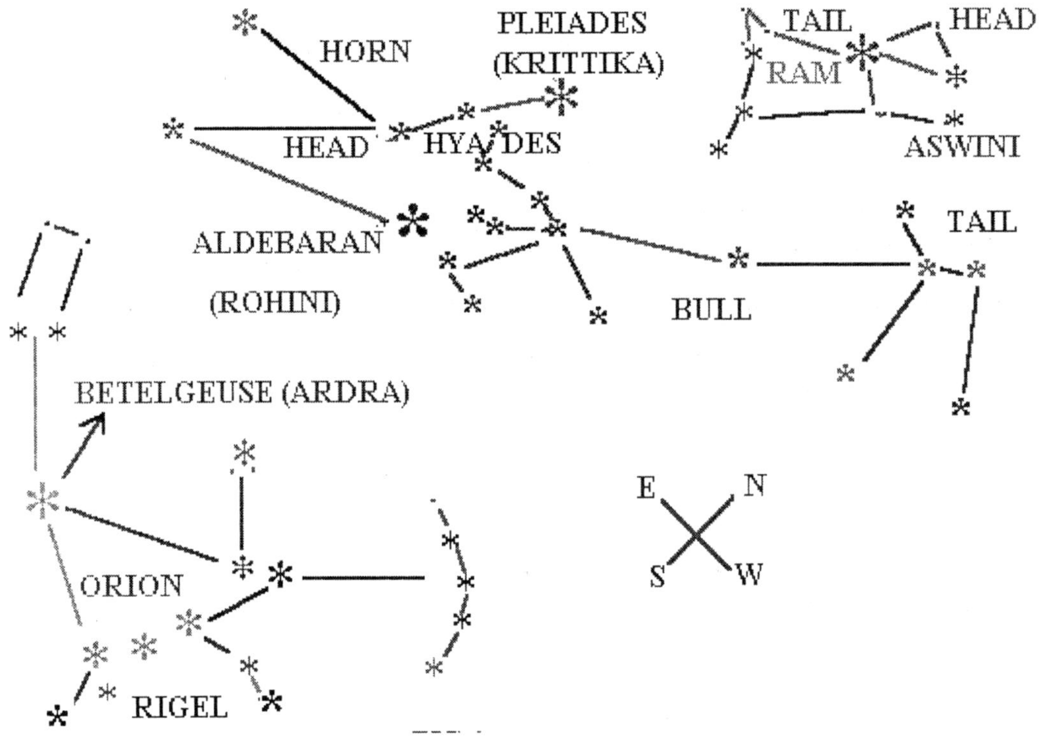

FIG 1

2. None of the star maps here and elsewhere in this book are drawn to scale. They are only indicative of the mutual locations.

Rohinî: This is a red giant, known as **Aldebaran**. It has a diameter 36 times as large as the Sun's, 100 times as luminous and 55 light-years away. This means that light takes 55 years to reach us from Aldebaran whereas it takes only 8 minutes to reach us from the Sun (at the same speed!). A red giant is a stage in the cosmic life of a star. Stars with diameters 10 to 100 times the diameter of the Sun are called **giant stars**; whereas stars with larger diameters are called **supergiants**. **Rohinî** can be seen near the meriaian around December l5 at 23 hours and accordingly at other times and nights according to the thumb rule table.

Mrigaśirah: This is λ-Orionis. Orion is a famous constellation with a number of very bright stars. See Fig. 2. We shall see more of this when we describe the next star. But before we do that it is interesting to refer to a legend in the *Satapatha Brahmana* of the *Yajur Veda*. It appears Lord Prajâpati fell in love with his own daughter Rohinî and in the form of a deer pursued her while the latter was running away in the form of a female deer. Enraged at this incestuous chase and prompted by the divines, Lord Rudra took the form of a hunter (*Mriga-vyâdha*) and pierced Prajâpati with his arrow. The head of the antelope with the arrow piercing it became the constellation Mrigaśirah.

Ârdrâ: This is the brightest star in the constellation Orion, which is the most prominent group of stars in wintertime. The three bright stars in a straight line form what is called **Orion's Belt**. The end stars in this belt measure an arc of 3°. Note that *one fist, thumb included, covers about 10 degrees of arc in the sky*. Orion is a hunter by profession, heavily armed, with a raised club (what is named *Mrigaśirâh* by Hindu astronomy). On Orion's shoulder is the reddish supergiant, **Betelgeuse**. This is ***Ârdrâ***. Diagonally opposite to this across Orion's Belt is Rigel, another giant star. It is in fact so much more luminous than Betelgeuse, that even though it is farther away and of smaller size, it appears almost of the same brightness to us. These two brilliant stars, along with Aldebaran (= *Rohinî*) and Sirius (*Mrigavyâdha*), form a very conspicuous lozenge of first magnitude stars. A line drawn through the three stars of the belt in each direction will have Aldebaran at its northwestern and Sirius at its southeastern end. Incidentally Sirius with magnitude −1.46 is the brightest star of the night sky. It is one of our closest neighbours among the stars. It is only 8½ light-years away from us.

Punarvasu: This is a pair of very bright stars, namely α and β-**Geminorum**. Their popular names are: **Castor and Pollux**. Pollux is the brighter star though the other one is called α. In fact these two are supposed to be the heads of the Twins which gives the name Mithuna for the constellation. Though these two are coupled in all first descriptions of the night sky, these two stars Castor and Pollux are very much different from each other. Castor is itself a multiple star of which the major components are two blue stars with magnitudes 2.0 and 2.9 respectively. Pollux is quite an ordinary star. But the location of these two bright stars in the Zodiac gives us a very beautiful show whenever the moon and perhaps a planet also are passing through Gemini. December through May is the best time to see ***Punarvasu***.

Pushya: This is actually a cluster of stars known as Praesepe, also called the Beehive Cluster. It covers more than 80 minutes of arc, an area almost three times the diameter of

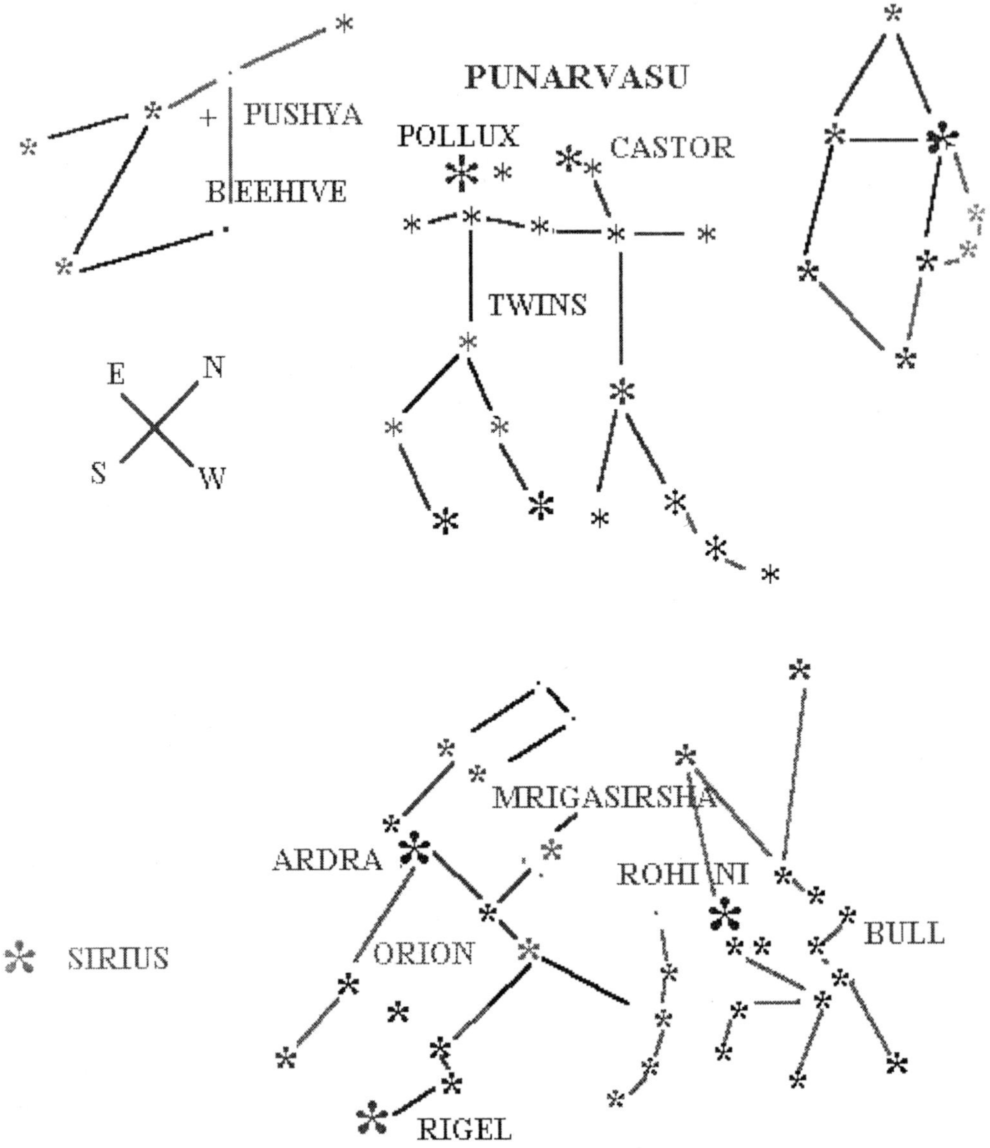

FIG 2

the moon. It is one of the most spectacular clusters in the sky just visible without glasses as a hazy spot under best conditions. It is almost right on the ecliptic. Look for this around March 1 at 10 p.m right above you.

Aśleshâ: Within only half an hour of longitude distance from ***Pushya***, we can see four stars, a little to the South of it and a little east of a bright star (mag.0.38) named Procyon. This bright star which is actually yellow-white forms almost an equilateral triangle with Sirius and Betelgeuse. It is the eighth brightest star in the sky and one of the closest stars to the Earth. The asterism Âsleshâ is made up of the four stars identified as ε, ρ, ζ, δ - **Hydra**. An imaginary line joining Castor and Pollux will cut the ecliptic to the west of Pushya and go into the middle region of *Âsleshâ*. Incidentally, the six stars all of them bright (magnitudes varying from 1.38 to –1.46), **Aldebaran, Rigel, Sirius, Procyon, Pollux,** and **Capella** form a large hexagon in the winter sky which cannot be missed. The red giant Betelgeuse is inside this hexagon, though not in its dead centre. **Capella** is the sixth brightest star in the sky.

Maghâ: In the middle of April, around 9 p.m. we can see this bright star Maghâ (=**Regulus** = α-**Leonis**) almost in the centre of the sky (for a 25° north latitude observer), with two brighter stars **Arcturus** further east and Spica in the South, the three forming a big isosceles triangle with Regulus as the vertex. To locate Arcturus (=*Svâti*) one follows the twist of the tail of the *Saptarishi*, — it is the curve of the Big Dipper's handle — over about 30°. And that will end up in Arcturus. To locate *Maghâ* one follows the two stars which are not the pointers (that is, in the rear of the bowl) and follows them in the backward direction for about 35°. This ends up in Regulus. See Fig. 3. Another tradition holds that Maghâ is a group of 5 to 6 stars in the constellation Leo.

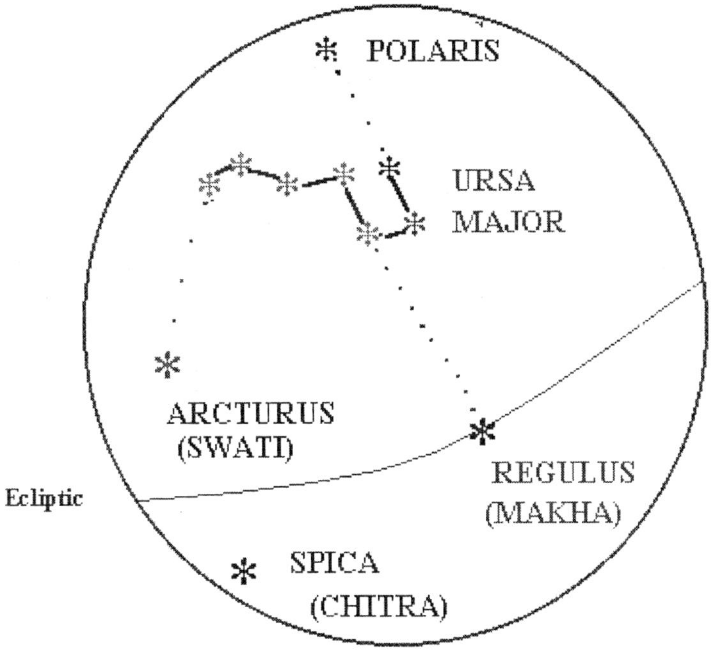

FIG 3(a)

Regulus has magnitude 1.35. It is a hot white star with surface temperature of about 14000°. It is 140 times as luminous as our Sun. It is ten times farther away than Sirius, the brightest star in the sky. If it were at the distance of Sirius, it will appear six times as bright as Sirius. Regulus is also a large star, with 2.8 times the diameter of the Sun. In fact Regulus is a single physical system made up of three totally unlike stars, one of two which is like our Sun, and the other is a white dwarf.

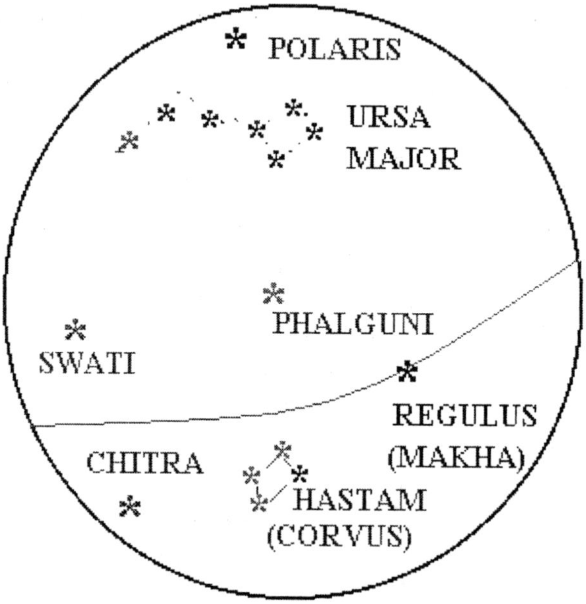

FIG 3(b)

Pûrva-Phalgunî: This is θ – **Leonis**, of magnitude 3.3. It is known by the common name Chertan in Western Astronomy.

Uttara-Phalgunî: This is β – **Leonis** of magnitude 1.7. Also known by the name Denebola. This and the previous star form with two others a conspicuous rectangle just 15° east of Regulus.

Hastâ: This asterism comprises of five stars in **Corvus** in the form of a trapezium . These are δ, γ, β, ε, α – **Corvus**. Hastâ means 'hand'. The best times to see this are the months April, May, June and July. This constellation is known in the western world as the crow (Meaning of Corvus). The five stars that constitute this asterism are counted in Virgo by ancient Indian Astronomy and it is still so counted in Indian Astrology. But the international convention places it in Corvus.

Chitrâ: This is a 0.98 magnitude star in Virgo, and is the brightest star in that constellation, Known as **Spica**, it is the 16th brightest star in the sky. It is only five times as large across, as our Sun. It is almost on the ecliptic and so it is sometimes hidden by the passing moon. The four bright stars **Spica** (*Chitrâ*), **Arcturus** (*Svâti*), **Car Caroli** and **Denebola** form what is known as the **Virgin's Diamond**. Around the middle of May round about 11 p.m. one can see Spica just west of the meridian.in the southern part of the sky.

Svâti: This is **Arcturus**, the fourth brightest star in the sky. Its magnitude is 0.0. Its technical name is α – Bootis. It is 36 light years away from us and has 25 times the diameter of the Sun. It is a giant star 100 times as luminous as our Sun. It is the first star one sees in late spring and early summer, after sunset high up in the northern sky. If we look at the tail of the Saptarishi and follow its sweep over an arc of 30° that is, a width of three fists including the thumb, we come to Arcturus as we have already observed in the description of Maghâ .

Viśâkhâ: This is in the zodiacal constellation Libra. This constellation has no bright stars and the shape of a pair of scales (*Tulâ*) is difficult to recognise. Even the brightest star has only a magnitude 2.9. This and five other stars make up what is known as the asterism Viśâkhâ in Hindu tradition. One of the end stars in Tulâ has a faint green colour. Probably this is the only green naked eye star in the sky.

Anurâdhâ: The largest easily recognisable constellation in the southern skies is Scorpion. See Fig. 4. The six stars at the head end of the Scorpion form *Anurâdhâ*. The best time to see this is the months of May, June, July August.

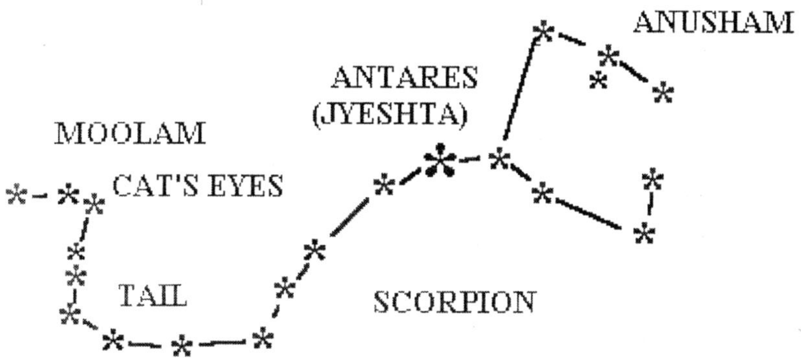

FIG 4

Jyeshthâ: This is a brilliant, distinctly reddish, first magnitude star called **Antares**. It is the 15th brightest star in the sky (magnitude 0.8) and the brightest star in Scorpio. It is a super-giant. It has a diameter 300 times as big as the Sun and 3000 times as luminous. It is its distance from us (300 light years) that makes it just a first magnitude star. It is close to the ecliptic and so is sometimes hidden by a planet like Mars, Jupiter or Saturn passing through. When Mars is very near to it it is sometimes difficult to recognise which is Antares and which is Mars because both are red. Of course we have to look for the twinkling of the star. Mars shines with a calm steady light.

Mûla: Almost at the tail end of the scorpion, it is one, the brighter one, of a close pair of stars called 'cat's eyes'. Magnitude 1.6. It is also called Shaula.

Pûrvashâdhâ and Uttarashâdhâ: Together these two asterisms constitute a quadrilateral of four stars in Sagittarius. One tradition holds that these are τ, σ, φ, ε – Sagittarium. Another tradition holds these are δ, ε, σ, ζ – **Sagittarium**. This area is full of beautiful star clouds in the centre of the Milky Way. The farther south you are the higher in the sky does Sagittarius rise and the sight is more fantastic. Sagittarius comes to the meridian at 22 hours on August 1 and on other dates and times according to our thumb-rule table.

Šravana or Šrona: This is Altair, the 12th brightest star in the sky. Its technical name is α – **Aquila**. It can be identified on September 15 on the meridian in the northern sky at 21-30 hours (and at other times accordingly) as a bright star chaperoned by two faint stars on either side. This is one reason that the Tamil description of this star says it is a triad of stars (see Chapter 4).

Šravishthâ or Dhanishtâ: Further east of Altair there is a small quadrilateral of stars which are so close to one another that it is very difficult to miss it on clear dark nights. The brightest star among them is α – **Delphinus** also called **Sualicin**. This is Šravishthâ. It can be seen in the meridian at 22 hours on September 1.

Šatabhishaj: This is a Y-shaped set of four stars in Aquarius namely π, ζ, η, γ – Aquarium. Another tradition says that it is the entire thickly studded cluster including the above four stars. There is also another tradition that it is just λ – Aquarium. This latter is a little east of the former. The Y-shaped asterism is also caslled the 'water jar' which gives the name to the constellation Kumbha. The star at the centre of the Y is described as the finest double star in the sky with components of magnitude 4.3 and 4.5. This comes to the meridian on October 1 at 21:30 hours.

Pûrva-proshtapadâ and Uttara-proshtapadâ: We come now to the **Great Square of Pegasus** which can be easily recognised in the southern sky as a large square shown by its four vertices, in the months of August, September, October and November - in the eastern half of the sky in the first two months and in the western half of the sky in the latter two months. The two bright stars among them form one west-to-east side of the square on its northern side. They are β – **Pegasus** (a red giant) and α – **Pegasus** respectively. They are our stars *Pûrva-proshtapadâ and Uttara-proshtaspadâ* See Fig. 5. It is supposed to be a test of good eye-sight to count, on a moonless cloudless night, as many as 25 stars within the square of Pegasus with the naked eye!

Revatî: Directly south of the square of Pegasus a boat shaped set of five stars constitute *Revatî*. It can also be imagined as a small irregular pentagon. It is called the 'circlet'. Actually the ecliptic passes through it. The vernal equinox γ is only 5° west of this.

Appendix II gives a sequence of 27 charts, that may be called Ready Reckoner for the Visibility of the 27 Nakshatras. Each of these charts gives, for the Nakshatra under consideration, the approximate dates of the year and the portion of the sky (Eastern or Western) where one can look for them at the various times of the night on these dates.

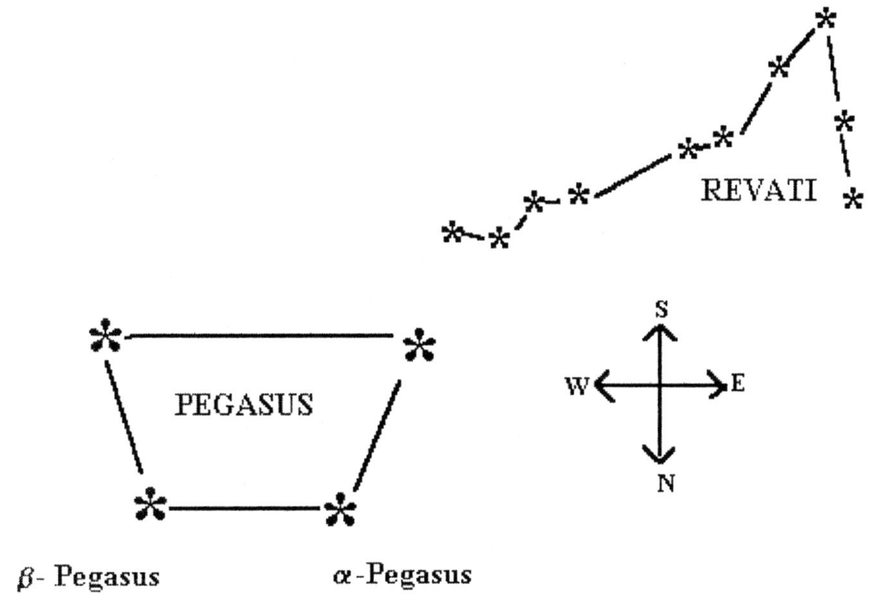

FIG 5

Chapter 3

The 27 Sanskrit Formulae

The 27 Sanskrit formulae which form the subject matter of this chapter can be learnt by rote in a day. Their purpose is the following. By just one look at the transiting stars or asterisms at any point of the night one can immediately tell the local solar time of the night. The only datum that one would need to know is the current date in the solar year. The principle is just this. If you want to know what time of night it is, you should know how far in its path below the horizon the Sun has travelled. But the calendar date is available. This means that the longitude of the Sun, i.e., the distance from γ on the ecliptic the Sun has travelled, is known. The formula then tells you: If you see such and such an asterism (star or group of stars) on the meridian, then such and such portion of the ecliptic is rising. To make this latter precise, the formula says that such and such zodiacal constellation is rising and it also tells you how much it has risen. Each formula has three words in it. The first word names the asterism that is transiting. The second names the sign of the Zodiac that is rising simultaneously. How much it has risen above the horizon is given by the third word which is a code for that number expressed in time units, the unit being one *Ghatika*, i.e., 24 minutes.

The four columns in the table on page 23 indicate as follows: Col. 1 gives the asterism number to be seen on the meridian. Col. 2 gives the zodiacal sign which is rising at the time. Col. 3 and Col. 4 give the amount by which that sign has risen. Col. 3 gives the whole number of Ghatikas (Each ghatika = 24 minutes), and Col.4 gives the additional fraction of a *Ghatika* as so many eighths and therefore equal to so many 'three minutes' and Col. 5 gives the no.of minutes corresponding to the given no. of *Ghatikas* and fraction of a *Ghatika*.

At this point an explanation regarding the numerical code is necessary. The standard usage is what is known as the *Ka-ta-pa-yâ- Sankhyâ*. This is the numeration which begins with the alphabet syllables Ka, Ta, Pa and Ya. It is a many-one correspondence between consonants of the Sanskrit alphabet and the numerals 1, 2, 3,.... 9 and 0. The notation goes back to Vararuci, one of the earliest Vedic commentators, and has been in extensive use in all of Indian literature, religious as well as secular. The table on page 24 gives the correspondence in full.

THE 27 FORMULAE OF THE CLOCK OF THE SKY

(as per the tradition of the author)

FORMULAE IN SANSKRIT	ENGLISH EQUIVALENT				
	COL. 1	COL. 2	COL. 3	COL. 4	COL. 5
Ashvi Karkî Roopâ	1	Cancer	2	1	51
Bharani Karkî Šâpâ	2	Cancer	5	1	123
Krittikâ Simhe Kâyâ	3	Leo	1	1	27
Rohinî Simhe Gurû	4	Leo	3	2	78
Mrigaširâh Kanyâ Neyâ	5	Virgo	0	1	3
Ârdrâ Kanyâ Yutâ	6	Virgo	1	6	42
Punarvasu Tulâ Neelâ	7	Libra	0	3	9
Pushya Tulâ Râma	8	Libra	2	5	63
Âshleshâ Tulâ Bhogî	9	Libra	4	3	105
Maghâ Vrischika Kalâ	10	Scorpio	1	3	33
Phalguni Vrischika Latâ	11	Scorpio	3	6	90
Uttiram Dhanur Nâma	12	Sagittarius	0	5	15
Hastâ Dhanur Balâ	13	Sagittarius	3	3	81
Chitrâ Dhanuh Šaraih	14	Sagittarius	5	2	126
Swâti Makare Kâmâ	15	Capricornus	1	5	39
Vishâkâ Makare Vanam	16	Capricornus	4	0	96
Anûrâdhâh Kumbhe Nâbhih	17	Aquarius	0	4	12
Jyešthâ Kumbhe Kathâ	18	Aquarius	1	7	45
Mûlam Kumbhe Gaccha	19	Aquarius	3	7	93
Pûrvâshâdhâ Meene Kapih	20	Pisces	1	1	27
Uttarâshâdhâ Meene Gathâ	21	Pisces	3	7	93
Šronâ Meshe Ratâ	22	Aries	2	6	66
Šravishthâ Vrishabhe Nabhah	23	Taurus	0	4	12
Šatabhishaj Vrishabhe Râma	24	Taurus	2	5	63
Proshtavân Mithune Parah	25	Gemini	1	2	30
Ahirbudhne Mithune Bhânu	26	Gemini	4	0	96
Revatî Karkî Yagnah	27	Cancer	1	5	39

Ka -ta-pa-yâ Sankhyâ

ka	ṯa	pa	ya	....	1
kha	ṯha	pha	ra	....	2
ga	ḏa	ba	la	....	3
gha	ḏha	bha	va	....	4
gna	ṉa	ma	ša	....	5
cha	ta	—	sha	....	6
ccha	tha	—	sa	....	7
ja	da	—	ha	....	8
jha	dha	—	—	....	9
jna	na	—	—	....	0

The diacritical underlinings below the letters in the second column indicate that they are cerebrals, as distinct from the dentals which are down below. Ša and Sa are differently pronounced. *ka-ta-pa-yâ Sankhyâ* is so-called because the counting from 1 starts from each of the consonants ka, ta, pa and ya. Note that the same numeral is represented by more than one consonant. Also note that in the use of the *Sankhyâ*, only consonants matter. All vowels used are to be ignored. Incidentally, these two elements of freedom enable the user to introduce a good amount of poetry in his symbolism!

Once we recognise one of the 27 asterisms on the meridian, we apply the formula corresponding to that asterism. It tells you what sign of the Zodiac is rising and by how much it has risen. Using the known longitude of the Sun for the day we do a mental calculation and tell the time of night. The correspondence between the position of the Sun and the calendar date is given in the succeeding table.

Zodiac Signs and Corresponding Dates When the Sun is in That Sign

Zodiac Sign	Sanskrit Equivalent	Corresponding Dates
Aries	Mesha	April 17 to May 16
Taurus	Vrishaba	May 17 to June 17
Gemini	Mithuna	June 18 to July 18
Cancer	Kataka	July 19 to Aug. 23
Leo	Simha	Aug. 24 to Sept. 19
Virgo	Kanya	Sept. 20 to Oct. 19
Libra	Tula	Oct. 20 to Nov. 18
Scorpio	Vrischika	Nov. 19 to Dec. 18
Sagittarius	Dhanuh	Dec. 19 to Jan. 17
Capricornus	Makara	Jan. 18 to Feb. 17
Aquarius	Kumbha	Feb. 18 to March 17
Pisces	Meena	March 18 to April 16

For the purpose of what follows, it is important to understand the (apparent) movement of the Sun through the signs of the Zodiac a little more in depth. The Sun takes roughly one month to move through one sign. The twelve signs have a span of 360° in the sky. So one sign of the Zodiac occupies 30° span of the sky. Every day, because of diurnal motion, every celestial object moves through 360° in about 24 hours. So a 30 degree-span of the ecliptic (apparent annual path of the Sun) will take roughly two hours to cross the same point (in its apparent daily procession across the sky), for instance. its highest point.

A visual help can be had in the form of the following two rough sketches. One shows the annual path of the Sun (now drawn as a circle in the celestial sphere).The dates when the Sun transits from one sign (of the Zodiac) to another are mentioned and the signs are written in a counterclockwise order (from the western horizon to the eastern horizon). The other sketch shows the nightly passage (not visible) of the Sun below the horizon from 6 P.M. to 6 A.M.

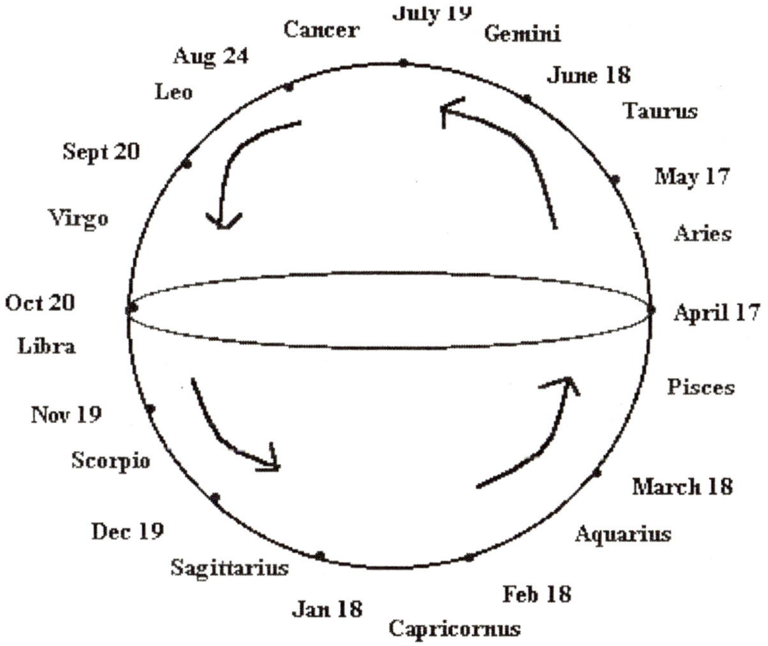

SKETCH 1

We shall now illustrate the application of these formulae by a succession of seven examples. In each case a figure is drawn to help the visualization. But the figure is only an approximation since it shows only a circle, to be taken as the circular path of the Sun (namely the ecliptic) on which are displayed some of the Zodiacal signs necessary for the calculation. The appropriate star is also shown located on this circle in its own constellation. The same circle serves also to indicate the diurnal path of the Sun on that particular day, since the ecliptic glides on the equator each day as the earth revolves around its axis. The eastern and western points of the horizon are as noted in sketch 2.

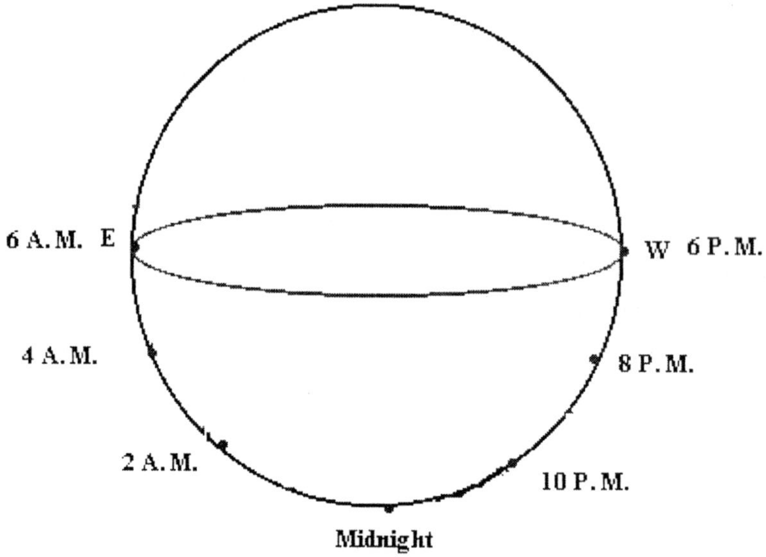

SKETCH 2

Example 1. Take the formula *Jyeshtâ Kumbhe Kathâ*. This means that when Jyeshtâ (that is, the star Antares) is on the meridian, the sign Kumbha (that is, the sign Aquarius) is rising on the eastern horizon to the extent indicated by the word Kathâ. This last word in Ka-ta-pa-yâ Sankhyâ means 'one and seven'. Together the word is equivalent, in the convention of these formulae, to a time duration of one ghatikâ and seven-eighths of a ghatikâ. This counts up to a duration of 24 + 21 minutes, that is, 45 minutes. Thus Kumbha has risen 45 minutes. The portion of Kumbha that is below the horizon is 1^h 15^m. Now comes into play the date of observation in question. Suppose the date on which this observation is made is April 8. On this date the Sun, in its annual motion, is in that position of the ecliptic where it has 8 more days to go in the zodiacal sign Meena (Pisces). Recall from sketch 1 above that the Sun, in its annual motion, goes from Kumbha to Meena and from Meena to Mesha and so on. And each day means 4 minutes of space in the diurnal motion. Since 8 days means 32 minutes of time, the Sun has already covered 1^h 28^m of Meena-space. See now figure 1 (a). On the day in question the time-difference between the eastern point of the horizon and the position of the Sun (at the time of observation) is

1^h15^m in Kumbha *plus* 1^h28^m in Meena.

Thus the time is 2^h 43^m before sunrise. In other words, the time is 3-17 A.M. , where for comvenience we take sunrise to be at 6-00 hours.

Example 2. Now we take the formula *Hastâ Dhanur Balâ*. Suppose we observe the star *Hastâ* (Corvus) on the meridian, say on May 1. The code Balâ is equivalent to 81 minutes and so the sign Dhanur has risen 1^h 21^m. On May 1 the Sun is in the middle of Mesha (Aries) which means it has covered one hour of Aries. See Fig. 2 (a). Thus the time

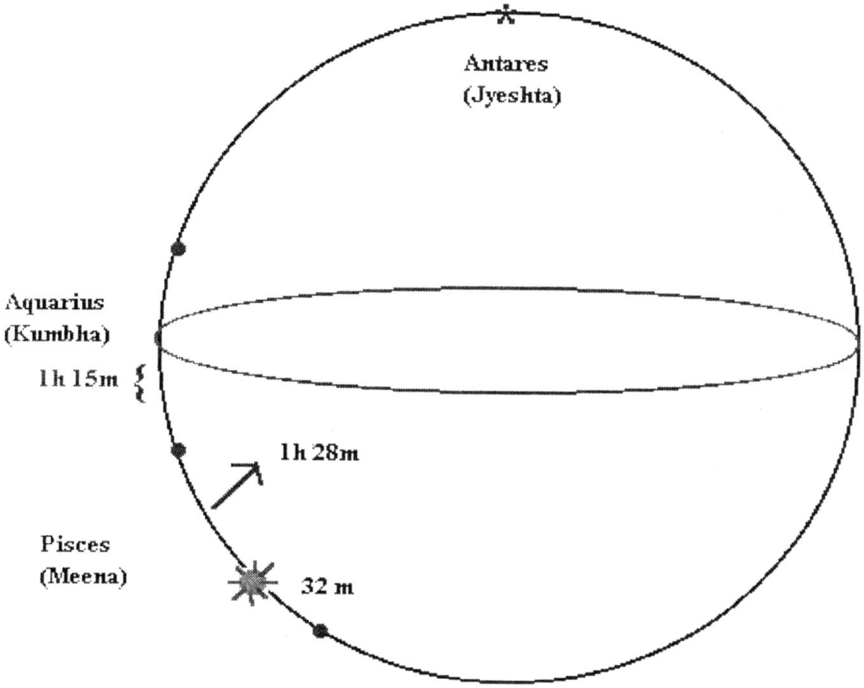

JYESHTAA KUMBHE KATHA
FIG 1(a)

difference between the eastern point of the horizon and the position of the Sun at this time of observation is

39^m in Dhanur *plus* 2^h of Makara *plus* 2^h of Kumbha *plus* 2^h of Meena plus 1^h of Mesha

Thus the time is 7^h 39^m before sunrise. In other words it is 10-21 P.M.

In both these examples we located according to the formulae what sign and what position of that sign was on the eastern point of the horizon and calculated the time difference between this position and the position of the Sun as of that date. It is very often more convenient to draw or imagine a figure corresponding to sunrise or sunset as the case may be. **Count from the Zodiacal sign indicated by the formula up to the sign where the Sun is on that date. If this number is less than or equal to 3, draw the sunrise figure. If the number is greater than 3, draw the sunset figure.** The wisdom of this recipe will be seen by actual experience.

Now take Example 1 again. Kumbha to Meena gives the number 2. So we draw the sunrise figure. See figure 1 (b) on page 29. At sunrise since there are 8 days for the

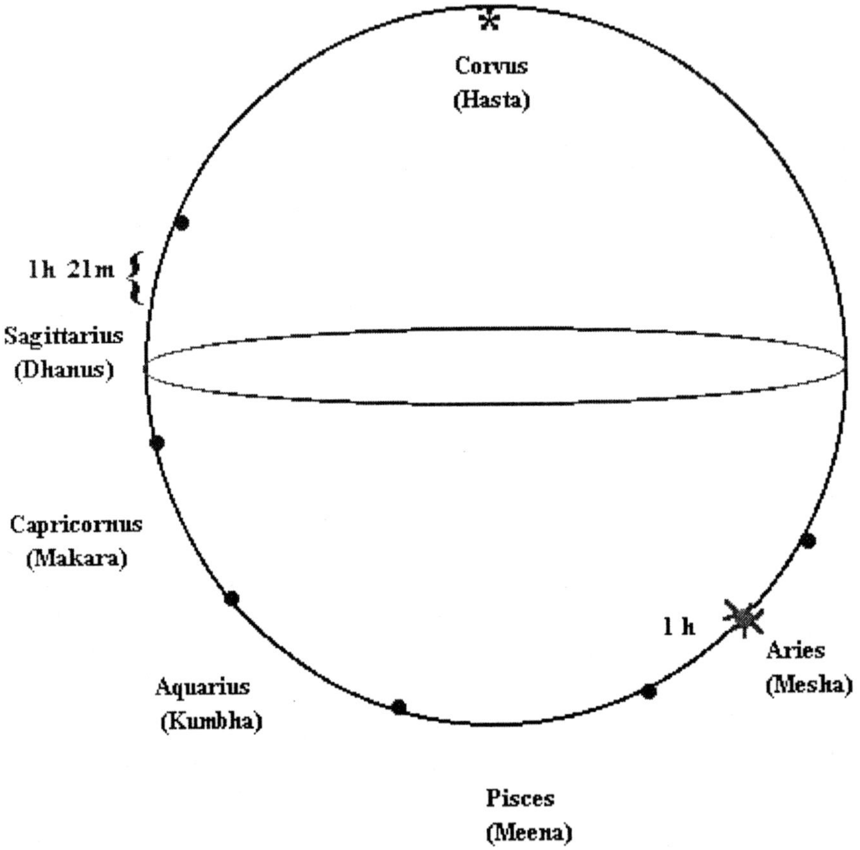

HASTA DHANUR BALA
FIG 2(a)

Sun to go in Meena there is 32 minutes of time space below the Sun (which is at the eastern point of the horizon). Therefore there is 1^h 28^m of Meena -space above the Sun and further above that 1^h 15^m of Kumbha-space in order to reach upto the point P indicated by the formula. Therefore, 2^h 43^m separates the point indicated by the formula and the rising Sun. In other words if P were at the eastern point of the horizon the Sun will be 2^h 43^m below it. Therefore the time is 2^h 43^m *before sunrise*.

In the case of Example 2 the count from Dhanur to Mesha is 5 and so we draw the sunset figure. The Sun is in the middle of Mesha. So at sunset one hour of Mesha is above the western point of horizon. This means the seventh sign from Mesha, which is Tula (See Appendix III) has one hour of its span below the eastern horizon. Below this there is two hours of Vrischika-space and below this there is $1^h\ 21^m$ of Dhanur-space, in order to reach the point P indicated by the formula. The time difference between the point P and the

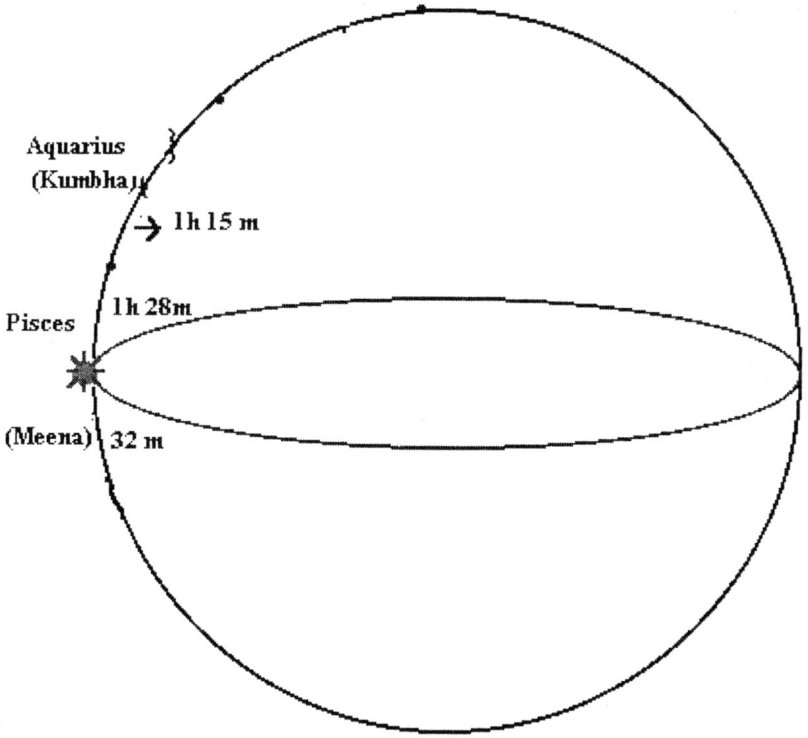

JYESHTAA KUMBHE KATHA
FIG 1(b)

eastern horizon is therefore 1^h of Tulâ *plus* 2^h of Vrischika *plus* $1^h\ 21^m$ of Dhanur, which totals upto $4^h\ 21^m$. This much time the Sun has to go down below the western horizon (see Fig. 2 (b) on page 30) in order for the point P to come to the eastern horizon. Thus the time of observation is $4^h\ 21^m$ *after sunset*.

Example 3. Let the date be January 7 and the star seen on the meridian be Krittikâ (Pleides). The formula is *Krittikâ Simhe Kâyâ*. The Sun is in Dhanur with 10 days to go. The count from Simha to Dhanur is 5. So we draw the sunset figure. See fig 3. There is 10 x 4 = 40 minutes of time space in Dhanus above the western horizon at sunset. Accordingly there is 40 minutes of Mithuna-space (Note that the seventh sign from Dhanur is Mithuna) below the eastern horizon. Below it there is 2^h of Karka-space and further below it there is 27m of Simha-space up to the point P indicated by the formula. Note that we have already translated Kâyâ into 24 + 3 = 27 minutes. Thus the time difference between the eastern horizon and the point P is $40^m + 2^h + 27^m = 3^h\ 07^m$. This much of time the Sun has to go in the western horizon for the point P to reach the eastern horizon. In other words the time is 9-07 P.M.

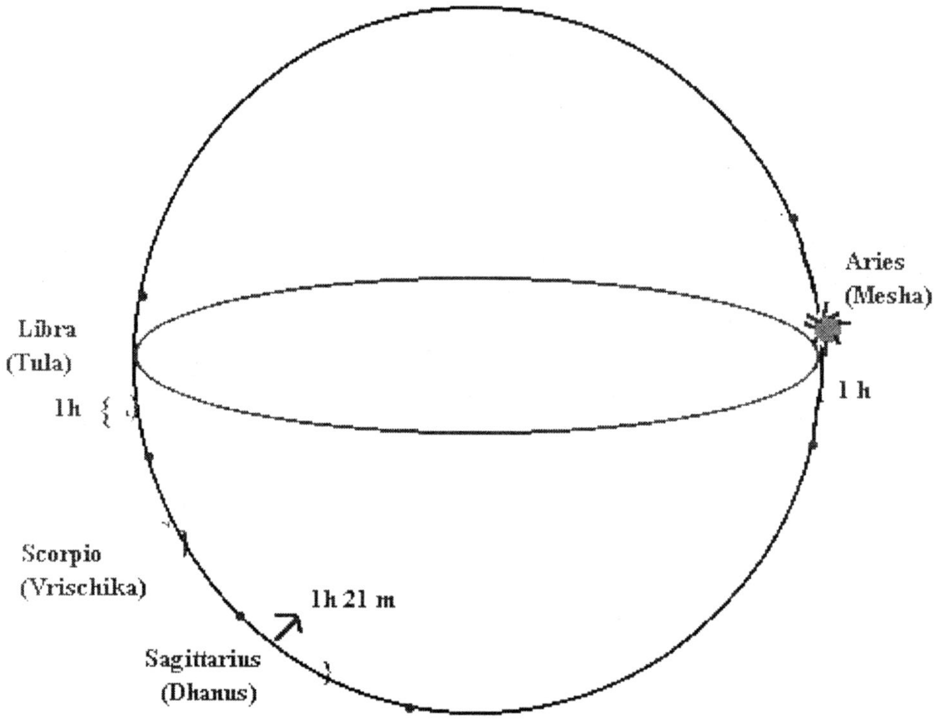

HASTA DHANUR BALA
FIG 2(b)

Example 4: Date is July 24 and the star seen is Uttara-Proshtapada (α-Pegasus). On July 24 the Sun is in Karka. The point P indicated by the formula *Ahirbudhne Mithune Bhanu* is in *Mithuna*. Count from Mithuna to Karka is 2. So we draw the sunrise figure. See Fig. 4. Since the Sun has already spent 5 days in Karka on July 24, the portion above the horizon in Karka is 20^m. Above that there is Mithuna portion. To calculate the time space upto the point P in the Mithuna portion we proceed as follows. *Bhanu* in our code means 4 x 24 = 96 minutes. This much is the portion of Mithuna above the horizon at the time of observation. So at sunrise the time difference between the Sun and the point P above it is 20^m plus (2^h minus 1^h 36^m) = 44 minutes i.e. the time of observation is 44^m before sunrise which means the time is 5-16 A.M.

Example 5. March 1. The Sun is in the middle of Kumbha with 16 days to go. Suppose we see Pushya on the meridian. Pushya, remember, is Praesepe cluster. *Pushya Tulâ Râma* is the formula. Counting from Tulâ to Kumbha we get 5. So we draw the sunset figure. See fig 5 below. 16 x 4 = 64 minutes. This much of Kumbha-space is above the western horizon. The seventh sign from Kumbha is Simha. So the time difference between the eastern horizon and the point P indicated by the formula is 64^m of Simha plus 2^h of Kanyâ plus *Râma* amount of Tulâ. This last one is 48^m *plus* 15^m = 1^h 03^m. So the required time difference in the eastern sky between point P and the eastern point of the horizon is

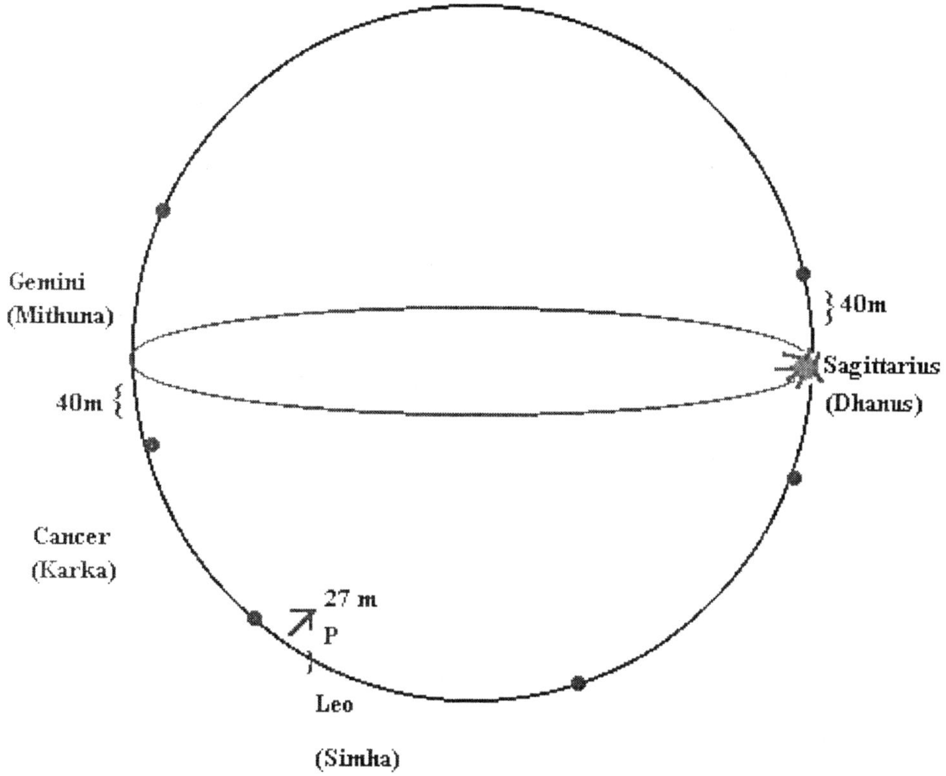

KRITTIKAA SIMHE KAAYA

FIG 3

$4^h\ 07^m$. This much the Sun has to go down in the western horizon for point P to reach the eastern horizon. The time of observation is therefore $4^h\ 07^m$ after sunset. In other words the time is 10-07 P.M.

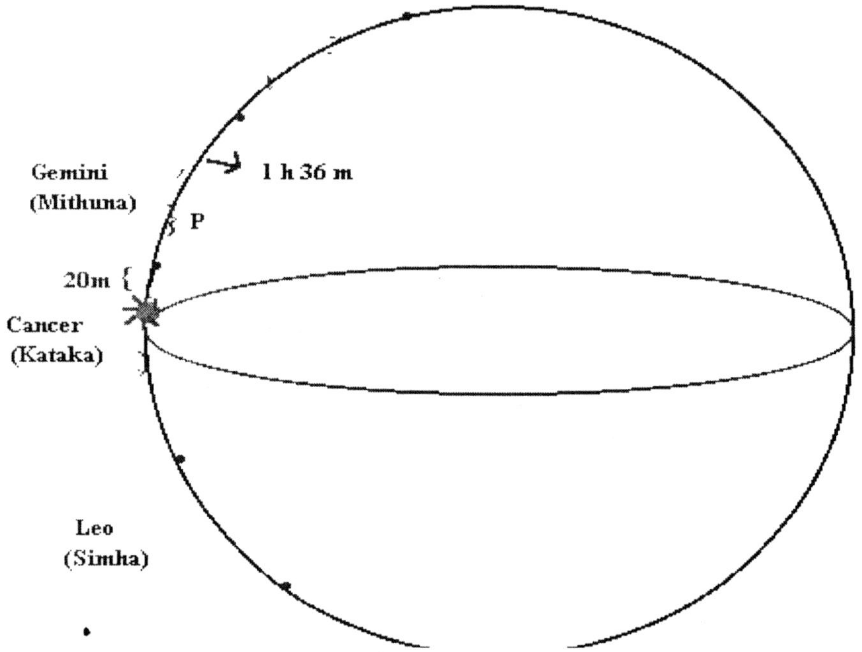

AHIRBUDHNE MITHUNE BHANU
FIG 4

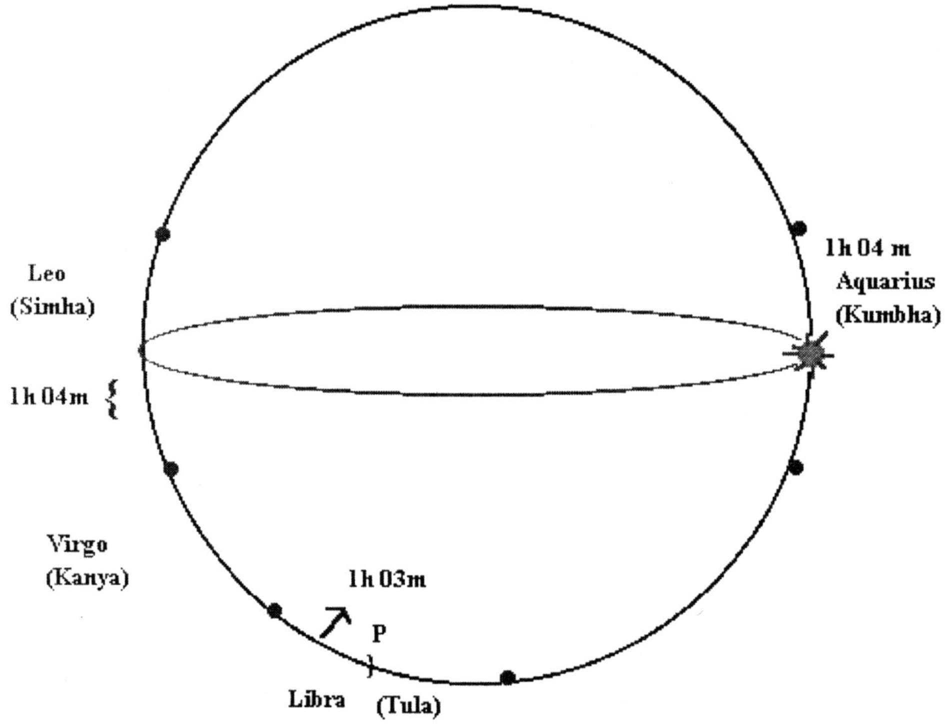

PUSHYA TULAA RAAMA
FIG 5

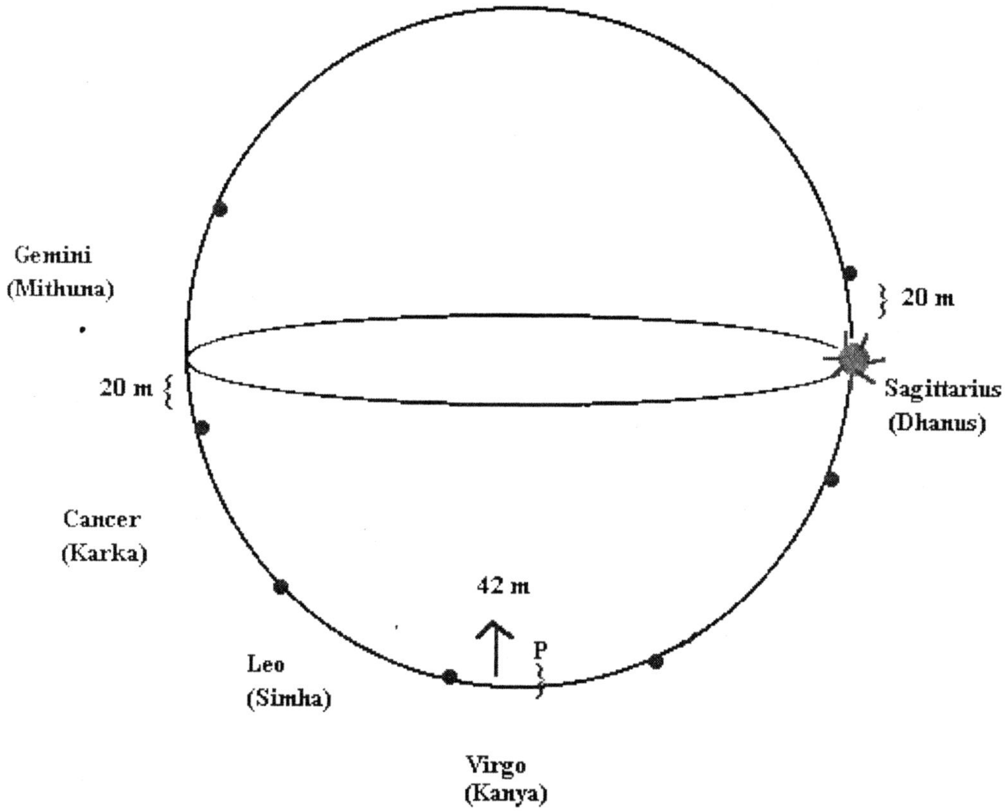

AARDRA KANYAA YUTAA
FIG 6

Example 6. Ârdrâ (Betelgeuse) is seen on January 12 on the meridian, say. The Sun is in Dhanur and the formula indicates Kanyâ. The count from Kanyâ to Dhanur is 4. So we imagine the sunset situation (see Fig. 6). The seventh sign from Dhanur is Mithuna. On January 12 the Sun has 5 days to go in Dhanur. So 20^m of Dhanur-space is above the western horizon and accordingly 20^m of Mithuna-space is below the eastern horizon. Then there is space equivalent to 2 hours of Karka, 2 hours of Simha and the *Yutâ* (recall the formula *Ârdrâ Kanyâ Yutâ*) amount of Kanyâ. This last one is 42 minutes. So the total time difference is $5^h\ 02^m$. This much the Sun has to go down in the western horizon for the point P to reach the eastern horizon. So the time is 11-02 P.M.

Example 7. Instead of having one more routine example we shall see now an unusual example which applies to a daytime situation. On October 24 of 1995 there was a total Solar eclipse round about 8-00 A.M. in the northern part of India across a small belt passing through Rajasthan, U.P. and Bengal. At this time the star *Magha* (Regulus) must have been seen near the meridian. The corresponding formula is *Magha Vrischika Kalâ*. When Magha is on the meridian Vrischika has risen 24 + 9 = 33 minutes. Draw the sunrise figure for this situation (see fig 7). On October 24 four days are over for the Sun in Tulâ. So at

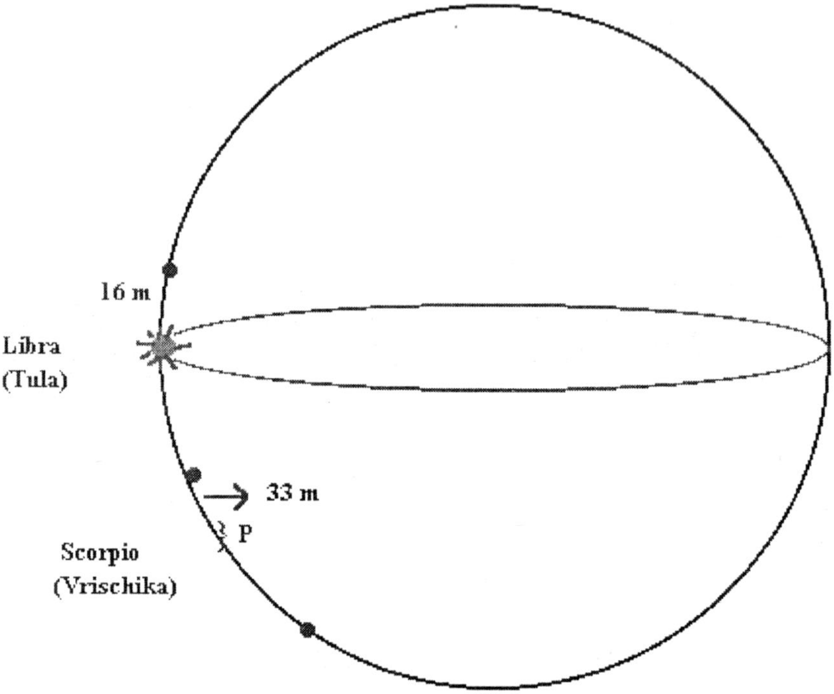

MAGHA VRISCHIKA KALAA

FIG 7

sunrise the time-difference between the Sun and the point P is $1^h\ 44^m$ plus $33^m = 2^h\ 17^m$. The Sun has to rise so much distance to get P on the eastern horizon. So when Magha is on the meridian the time is 8-17 A.M. Seventeen minutes before this time the star Magha would have been 17 minutes east of the meridian and exactly at that time the totality of the eclipse happened somewhere in this belt. So at that time of totality of solar eclipse the star Magha must have been visible, in that part of the belt of totality in India.

Before concluding this chapter let us note how the ancients applied a correction to these formulae. The reason for this correction is not very clear. It looks something like a mid-course correction. The correction goes in the form of a Sanskrit verse. This verse also was learnt by the author from his father.

> ***Jyeshtädhinäm atikramya Revatyädhin anägame***
> ***Ärdrädhimscha samam drishtvä nava-shat-dvädasa-kramäh.***

Literally this is what the verse says. The nine asterisms starting from Jyeshtâ must be seen *past* the meridian; the six asterisms starting from Revati must be seen *before* they cross the meridian; and the twelve stars starting from Ârdrâ must be seen exactly when they are *on* the meridian — for the purpose of the above formulae!

In the above examples therefore, the asterisms have to be not necessarily seen exactly at the time of their upper transit. Some have to be seen a little earlier and some others a little later. A first conjecture thsat comes to one's mind is perhaps these corrections take account of the fact that some stars are nearer to the equator and some others are farther away as shown by their declinations. But this conjecture falls flat on a closer examination. However we shall come back to the matter of these corrections in our last chapter, where we critically examine these formulae.

Chapter 4

The Tamil Tradition

It is amazing that the Tamil tradition has two sets of formulae — one of them equivalent to the one in Sanskrit and the other a formula in one single verse. This latter is algorithmic in its nature unlike the Sanskrit formulae or the other Tamil formula. We shall take this up first. We should warn the reader that this is a very approximate formula; its beauty lies in the fastness with which it is able to arrive at the result.

The statement of the single-verse formula in original Tamil is the following:

> ***Chittiraikkup-pûsamudal seerâvanikkanushamâm***
> ***Attanusuk-kuttirattâdiyâm; nittam nittam***
> ***eyducham-ânâlum irandey-kâlir-perukki***
> ***mâdam-aindu talli madi.***

The formula contained in this cryptic verse may be called by the equally cryptic name 'Magic Tamil Formula' (MTF). A literal translation of the verse may be given as follows.

> **For Chittirai (count) from Pûsam; for Âvani, from Anusham; for Dhanus (=Mârgazhi) from Uttirattâdi; on any day whatever is in the meridian, multiply (the count) by two and a quarter and subtract five for every month.**

We shall explain all this now. The beauty of the formula is three-fold. First the entire calculation for all the stars is contained in one four-line verse. Secondly, what it contains is a simple algorithm. And thirdly, instead of being a thumb-rule formula to be uniformly applied for all time throughout the year it has an in-built correction factor for the varying months over the year, though, the varying situations within the month are not protected or compensated — and this is what gives it its approximate character.

First we have to know by rote all the 27 asterisms by name and in their natural order. This was given already at the beginning of Chapter 2. Secondly we have to know the names of the twelve Tamil months, which are actually the solar months of the year. These are:

Chittirai, Vaikâsi, Âni, Âdi, Âvani, Purattâsi, Aippasi, Kârtigai, Mârgazhi, Thai, Mâsi, Panguni.

The corresponding dates in the English calendar, are given in the following table. The dates have an error of plus or minus one even according to present usage. But they have a more serious error from the point of view of Astronomy. This is because of the precession of the equinoxes, which is not recognised by the Tamil calendar — however, this is an over-simplification. The Tamils do recognise the precession but only once in 2160 years when the equinox has moved back a full 30 degrees. Since it was last corrected to be around April 14, it has moved only around 24 degrees or so. The Tamils still work with April 14 as the beginning of their solar year. So the following table shows the dates against the Tamil months as per usage and not as per Astronomy. It is a good guess that this Tamil formula must have been concocted not earlier than the last correction (in the Tamil calendar) to the precession; in other words, not earlier than around 1750 years *ago*.

S. No	Tamil month	English dates
	First cycle for the purpose of the formula	
1	Chittirai	April14 to May 13
2	Vaikâsi	May14 to June 13
3	Âni	June14 to July 13
4	Âdi	July14 to August13
	Second cycle for the purpose of the formula	
1	Âvani	August14 to September13
2	Purattâsi	September14 to October13
3	Aippasi	October14 to November 13
4	Kârtigai	November14 to December13
	Third cycle for the purpose of the formula	
1	Mârgazhi	December14 to January13
2	Thai	January14 to February13
3	Mâsi	February14 to March14
4	Panguni	March15 to April 13

Now the meaning of the formula is this. Whatever asterism we see on the meridian, a count has to be made from one of three asterisms: *Pûsam* or *Anusham* or *Uttirattâdi*. Which one of these three stars it is, depends on what Tamil month it is in which you are making the observation. If it is the first cycle of four months it is *Pûsam*; if it is the second cycle of four months it is *Anusham*; if it is the third cycle , it is *Uttirattâdi*. Do this counting. If *n* is the number of counts multiply this *n* by 2¼ (that is, two and one fourth). From the result so obtained, subtract 5 *m* where m is the month number in the cycle in which it appears above. The resulting number is the number of ghatikas (units of 24 minutes) past sunset and therefore gives the time of night!.

Suppose it is May 18. And let the asterism be *Svâti* (Arcturus) that is seen on the meridian. May 18 is in the first cycle in the table above. So we have to count from Pûsam to Svâti. The count is 8. Multiplying this by 2¼ we get 18. The month number m corresponding to May 18 is 2. So we have to subtract 2 x 5 = 10 from 18. This results in 8. 8 *ghatikas* are equal to 3^h 12^m. So the time of observation is 9.12 p.m.

Just one more example. Suppose it is February 15. And let the asterism be Chitra (Spica) that is seen on the meridian. We have to count from Uttirattâdi. The count is 16. Multiplying this by 2¼ we get 36. The month number corresponding to February 15 is 3. We therefore subtract 3 x 5 = 15 from 36 thus arriving at 21 *ghatikas*. This is equivalent to 8^h 24^m. The time of observation is therefore 2-24 a.m.

Since the whole business is algorithmic we can build them into twelve tables, one for each month showing the answers for the particular asterism seen on the meridian. Below are three sample tables, one for each of three months selected from the three cycles above. See Appendix IV for the tables of the remaining months. Note that the minus answer indicates that the asterism reaches the meridian before sunset, that is, in daytime.

VAIKASI MONTH

STAR NAMES	**STAR NOS.**	COUNT FROM Pusam (N) (N)	NO. OF GHATIKAS PAST SUNSET [N x (9/4)] - 10
Pûsam	8	1	-7.75
Ayilyam	9	2	-5.5
Magham	10	3	-3.25
Pûram	11	4	-1
Uttiram	12	5	1.25
Hastam	13	6	3.5
Chittirai	14	7	5.75
Svâti	15	8	8
Vishakam	16	9	10.25
Anusham	17	10	12.5
Kettai	18	11	14.75

Moolam	19	12	17
Pürадam	20	13	19.25
Uttiradam	21	14	21.5
Tiruvonam	22	15	23.75
Avittam	23	16	26
Sadayam	24	17	28.25
Pürattadi	25	18	30.5

AIPPASI MONTH

STAR NAMES	**STAR NOS.**	COUNT FROM ANUSHAM (N)	NO.OF GHATIKAS PAST SUNSET [Nx (9/4)] - 15
Anusham	17	1	-12.75
Kettai	18	2	-10.5
Moolam	19	3	-8.25
Pûradam	20	4	-6
Uttiradam	21	5	-3.75
Tiruvonam	22	6	-1.5
Avittam	23	7	0.75
Sadayam	24	8	3
Pûrattadi	25	9	5.25
Uttirattadi	26	10	7.5
Revatî	27	11	9.75
Ašvini	1	12	12
Bharani	2	13	14.25
Kartigai	3	14	16.5
Rohinî	4	15	18.75
Mrigasirsham	5	16	21
Tiruvadarai	6	17	23.25
Punarpusam	7	18	25.5
Pûsam	8	19	27.75
Ayilyam	9	20	30

PANGUNI MONTH

STAR NAMES	STAR NOS.	COUNT FROM UTTHIRATTADI (N)	NO. OF GHATIKAS PAST SUNSET [N x (9/4)] - 20
Uttirattadi	26	1	-17.75
Revatî	27	2	-15.5
Ašvini	1	3	-13.25
Bharani	2	4	-11
Kartigai	3	5	-8.75
Rohinî	4	7	-4.25
Mrigasirsham	5	8	-2
Tiruvadarai	6	9	0.25
Punarpusam	7	10	2.5
Pûsam	8	11	4.75
Ayilyam	9	12	7
Magham	10	13	9.25
Püram	11	14	11.5
Uttiram	12	15	13.75
Hastam	13	16	16
Chittirai	14	17	18.25
Svâti	15	18	20.5
Vishakam	16	19	22.75
Anusham	17	20	25
Kettai	18	22	29.5

It will be interesting to see the 'Why' of this apparently magic-like but simple formula. First we have to appreciate the ingenuity of the formula in dividing the twelve months into three parts and devising three different starting asterisms for counting as well as different starting months fior the counting of months. This is based on the thumb-rule of equally dividing the 12 constellations of the zodiac for the 27 asterisms which is very much in vogue in the Indian science of Astrology and therefore in Astronomy also. This thumb-rule is not very much different from actuality. At least, for the purpose of calculation of this formula for knowing the time of night it works very well. The divisions of the zodiac into 27 parts is given in the following table. It shows how each constellation of the zodiac gets $2\frac{1}{4}$ of the asterisms. In practical calculations the space on the ecliptic of 13° 20' is used for each Nakshatra, which means that each of the 27 nakshatras gets an equal part of the ecliptic for itself.

	Asterisms	Constellation (Tamil name with English equivalent)
1	Aswini, Bharani, Kartigai ¼	Mesham (Aries)
2	Kartigai ¾, Rohini, Mrigasirsham ½	Vrishabham (Taurus)
3	Mrigasirsham ½, Tiruvadarai, Punarpusam ¾	Mithunam (Gemini)
4	Punarpusam ¼, Pûsam , Ayilyam	Katakam (Cancer)
5	Makham, Püram, Uttiram ¼	Simham (Leo)
6	Uttiram ¾ , Hastam, Chittirai ½	Kanni Virgo)
7	Chittirai ½ , Swati , Vishakam ¾	Tulam (Libra)
8	Vishakam ¼ , Anusham, Kettai	Vrischikam (Scorpio)
9	Moolam , Pûradam , Uttiradam ¼	Dhanusu (Sagittarius)
10	Uttiradam ¾ , Tiruvonam , Avittam ½	Makaram (Capricornus)
11	Avittam ½ , Sadayam , Pûrattadi ¾	Kumbham (Aquarius)
12	Pûrattadi ¼ , Uttirattadi , Revati	Meenam (Pisces)

In the month of *Chittirai*, at the beginning of the month, when the Sun is setting, it is the beginning of *Mesham* that is touching the western horizon. So the fourth constellation from *Mesham*, namely, *Katakam*, will have its beginning on the meridian. We can roughly take *Püsam* to be representative of *Katakam* because its full span is in the middle of Katakam. **Every asterism takes roughly $24^h/27$ hours to cross the meridian, that is, 54 minutes of time approximately, that is, 2¼ ghatikas.** So after Pûsam is on the meridian, the next asterism to be on the meridian is *Âyilyam*. The next one is Magham. When Magham is on the meridian, it is 3 x 2¼ that is 6¾ ghatikas after sunset. But wait. We want this formula to be applicable for a whole month, that is, till the Sun reaches the end of Mesham. This takes a span of 5 ghatikas in the daily motion. So the formula says, 6¾ minus 5 that is 1¾ ghatikas after sunset, you will see Magha on the meridian, spread as an average for the entire month of *Mesha*. There are several averaging processes here and so to that extent the formula is only a rough approximation. The three tables above and the remaining tables in Appendix IV show how the single formula takes care of the entire year and how the formula focusses on those stars that are visible in the night of the particular month of the observation, by showing the third column as a minus answer or as a number of ghatikas greater than 30. If the third column shows a minus answer it means the star will be on the meridian before sunset and so not visible. If the final answer is greater than 30 it means that the star comes to the meridian after sunrise and so is not visible. The only defect in the formula is that it treats all the days of the whole month the same way and in this sense, it is less accurate than the Sanskrit formula or the Tamil formula which we are going to describe now.

The 27 Tamil formulae which correspond to the Sanskrit formulae shall now be taken up for discussion. But mark it, these are not translations of the Sanskrit formulae. They seem to have an independent origin, though the style of the formulae is the same. In the following, for convenience in understanding, for the reader who may know some Tamil, the name of the Nakshatra (seen on the meridian) is underlined and also the zodiacal sign (which is rising at that time) is underlined.

The 27 Tamil Formulae

1. *Achwini arumïn kudiraittalaipol,*
   Mechidu kadaka ttiru kadikaiyatham.

2. *Bharani Mummïn aduppup pola,*
   Thiranmigu kadaka ttoru nârkâlâm.

3. *Kârtigai yarumîn etrrari yekam.*

4. *Urohini ûtrâl panniru mînâm,*
   Muranmigu Simham Münrekâlâm.

5. *Mânralai Mûnrum Teyngai kannpol,*
   Ânra singattainde kâle.

6. *Âdirai oru mani Âyizhai onrara.*

7. *Odaru punarmîn odangkolarai.*

8. *Pûsam pudalam pûvadu pola,*
   Vîsum kolil irandaraiyagum.

9. *Âyilyam nangum ammipola,*
   Mâyiru kolil nâle kâle.

10. *Makhamoru nângum; mudanugam pola,*
    Nagarum thelil onraraiyagum.

11,12 *Pûram , Uttiram, porrkarr kattil,*
    Ûrthell mûnre onnsilai yaraiye.

13. *Attam aimmin kaittalam pola,*
    Muttîn dhanuvin mûnrenappadume.

14. *Chittirai irumîn sirupulik kannpol,*
    Attanu vadanil aindenappadume.

15.	***Chodi tulakkattoru mani yenpa, Sâdhi makarattonrarai sâtrum.***
16.	***Visâkamärum idipâzhnduravâm, Visaiyaru makarattagu nârkadigai.***
17.	***Anushamârum mudappanai pola, Kuniyum kumbhat-taraiyâgumme.***
18.	***Kettai nangum Îtti pola, Ôttaik kudattil onraraiyâgum.***
19.	***Mûlamârum ekkâlam pola, Châlum kudattir chaduramâme.***
20,21	***Pûrâdam, Uttirâdam pûngâl nâlâm, Sîrâmînat-torarai nange.***
22.	***Ônam mûnrum uzhakkol pola, Kânum âttil mûnrenappadume.***
23,24	***Avittach chakkarattârâm erarai, Tavittuch Chadayantani yerirandarai.***
25,26	***Pûrattâdiyum , Uttirattadiyum, Ôrâmithunat tonrarai nange.***
27.	***Revathî moymmîn toani pola, Ovaru kadakat-torumukkale.***

One speciality of these Tamil formulae is that the formula itself contains a description of the star. This description is missing in the Sanskrit formula. But it has also to be said that the description is not of any great help in identifying the star, because it says very little about the star and almost nothing about the relative position of the star in the sky with respect to the other stars. But the startling fact about the skeletal description contained in the formula is that the formula talks as if it knows so much about the star — startling because, these formulae are pretty old, as old as the Sanskrit formulae themselves The succeeding table given below gives the 27 Tamil formulae of the clock of the night sky, the description of the stars and the English equivalent of the formulae.

STARS	DESCRIPTION	NOs.	RASI	NO. OF *GHATIKAS*	NO.OF *MINUTES*
AŚWINI	Six stars, like a horse's head	1	CANCER	2	48
BHARANI	Three stars like a country-oven, of shape ∩	2	CANCER	4	96
KRITTIKÂ	Six stars	3	LEO	1	24
ROHINI	Twelve stars	4	LEO	3¼	78
MRIGAŚIRAH	Like the three spots (eyes) on the coco-nut shell	5	LEO	5¼	126
ÂRDRÂ	Single star	6	VIRGO	1½	36
PUNARVASU	Boat-like	7	LIBRA	½	12
PUSHYA	Like the flowers of a snake-gourd	8	LIBRA	2½	60
ASHLESHA	Four stars like a block of stone	9	LIBRA	4¼	102
MAKHA	Four stars	10	SCORPIO	1½	36
PÛRAM	This and the next star together form a four-legged cot as it were	11	SCORPIO	3	72
UTTIRAM		12	SCORPIO	½	12
HASTAM	Five stars, forming a palm of the hand	13	SAGITTARIUS	3	72
CHITRA	Two stars like the staring eyes of a tiger	14	SAGITTARIUS	5	120
SWÂTI	Brightly shining single star	15	CAPRICORNUS	1½	36
VISHAKA	A congregation of six stars like a battered structure	16	CAPRICORNUS	4	96
ANUSHAM	Six stars —	17	AQUARIUS	½	12
JYESHTÂ	Four stars like a spear	18	AQUARIUS	1½	36
MOOLAM	Six stars —	19	AQUARIUS	4	96
PÛRADAM	This and the next star form a quadrilateral	20	PISCES	½	12
UTTIRADAM		21	PISCES	4	96
TIRUVONAM	Three stars like a small measuring glass	22	TAURUS	3	72
AVITTAM	Six stars forming a convex figure	23	TAURUS	½	12
SADAYAM	Single star	24	TAURUS	2½	60
PÛRATTADI	This and the next star form part of four	25	GEMINI	1½	36
UTTIRATTADI		26	GEMINI	4	96
REVATI	Like a boat	27	CANCER	¾	18

Take, for example, the formula no. 4. The formula says Rohini is made up of 12 stars. We have identified Rohinî as Aldebaran, α - Tauri. We know this is a red giant and a single star. So we are not able to understand what the formula means when it says that Rohinî is made up of 12 stars. Let us ignore the description and go to the second part of the formula. It says that when Rohini is on the meridian, Simha rises by 3¼ *ghatikas*. The corresponding Sanskrit formula also says that when when Rohinî rises, Simha rises by 3¼ *ghatikas*. In this sense the Samskrit and the Tamil formulae coincide. If the date, say is December 11, the Sun is almost at the end of Vrischika i.e. 7 days to go in Vrischika, i.e. Scorpio. So at the time of sunset, The Sun has crossed 28 minutes of space through Scorpio in the western horizon. The seventh constellation from it,Vrishaba is on the eastern horizon with 28 minutes below the horizon.

The position 3¼ghatikas (1^h 18^m) in Simha is 5^h 46^m below the eastern point of the horizon. So at the time of observation,the time is 23:46. See Fig. 1.

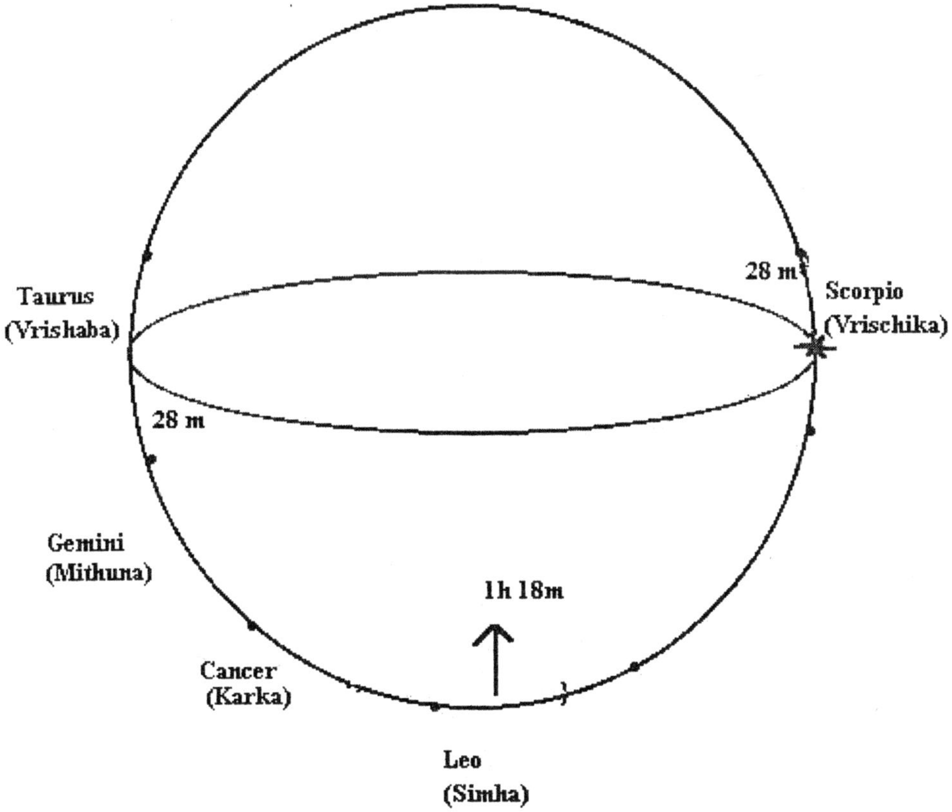

FIG 1

It will be clear that these Tamil formulae have been concocted in imitation of the Sanskrit formulae, except that they add a description of the asterism, which in most cases is only baffling. It is interesting to note that it talks of *Kârtigai* (Sanskrit *Krittikâ* corresponding to Pleides) as six stars. This is pleasantly surprising regarding the correctness of the description. But when it talks (in Formula No.10) about *Magha* (Regulus) as being of four stars, (in Formula No. 14) of Chittirai (Spica) being two stars, (in Formula No. 18) of *Kettai* (Jyeshta in Sanskrit, corresponding to Antares) as being four stars and (in Formula No. 22) of *Onam* (*Šravana* of Sanskrit, corresponding to Altair) as made up of three stars — we cannot but be baffled. We can only interpret these as descriptions of the area around or in the neighbourhood of the stars we are talking about. Which are the exact other stars which make up the asterism that is so named in Tamil is what is puzzling. The classical tendency in the world of Tamil literature (also in Sanskrit) to make every description poetical can be seen throughout these formulae. It says therefore Chittirai is made up of two stars looking like the two eyes of a tiger. We cannot agree that Chittirai is two stars and so there is no necessity to give any importance to the description about it being like the two eyes of a tiger. On the other hand, when Formula No.5 says that Mrigaširah (corresponding to Orion's Club) is three stars looking like the three weak spots in the shell of a coconut, it is interesting to identify them as the three stars one of which is λ-Orionis and the other two are very close to it, but less bright. This description is so nicely fitting that we cannot but admire the ingenuity of those who composed this description of the asterism. Similarly in the description of Hasta, which is five stars in Corvus, as the five stars looking like the spread of the palm of a hand, again we have nothing but admiration for the author. In view of these exact descriptions of these asterisms we may probably give a benefit of doubt to the author where the description is blatantly contradicting our identification of some other asterisms and only conclude that we do not have a full understanding of what is meant in the description.

Chapter 5

The Tanjore Manuscript

The previous four chapters have been written without referring to any written or printed source material. In fact no such material was in the author's hands when this book was started. In the case of the 27 Sanskrit formulae the only source was the verbal authority that the author got from his father who himself remembered it as traditional knowledge. In the case of the Tamil formulae described in Chapter 4, the author has no way of locating the source. In that sense this is also tradition, though of a different kind. Once the writing of this book got under way, the author consciously looked for some written evidence and contacted different scholars (in both Sanskrit and Tamil), astronomers, and astrologers, for a relevant reference in some ancient scientific or religious literature. It was during such a search that the author was fortunate enough to come across a Tanjore manuscript, located for him in the Tanjore *Saraswati Mahal Library* by Sri L. Narayana Sharma of Sanskrit College, Madras. This work is titled *Inakula-râja-tejo-nidhi* by *Tulaja-râja*[1], a Maratha King of Tanjore, a famous scholar and the author of several works including some on music including Sangîta-sâra-amrita, an extensive work on the Southern system of music. In the current work of his on astronomy and astrology there are twelve chapters. The eleventh chapter is called the Chapter on rising and setting (*Udaya-asta-adhikârah*). Just before this he mentions these formulae for finding the time of night by looking at the stars on the meridian. The formulae are a little different from what we have stated in Chapter 3, the essential differences being only in fractions of *ghatikâs*. There are actually two versions of the relevant pages in original Sanskrit. One version[2], which must have been the original written by Tulaja raja, uses Sanskrit characters for the numerals involved in the calculations. The other[3], which must have been a later copy of the same, has the numerals in Indo-Arabic characters.

---

1. Tulajarâja was the youngest son of Dîpâmbâ and Ekojî or Vyannkajî (Mahârâja 1675/1684), the son of Tukkabâi and Šâhaji Bhonsle (d. 1664), the son of Mâlojî (d. 1619), a noble in the service of the Nizâmshâhs of Ahmadnagar. Tulajarâja or Tukkkoji was the Mahârâja of Tanjore from 1728 to 1736.
2. By Courtesy of Indira Gandhi National Centre For Arts , New Delhi.
3. By Courtesy of Tanjore Saraswati Mahal Library, through Sri L. Narayana Sharma, Samskrit College, Madras

The interpretation of the Tanjore manuscript, however, is not so easy as the above paragraph might indicate. The reader may remember that the third word in each of our formulae of the 3rd chapter was interpreted according to a certain coding and a convention. The coding is the standard *Ka-ta-pa-yâ sankhyâ*. The convention is that the '*first letter represents the number of ghatikâs and the second letter represents the number of one-eighths of ghatikâs*'. This does *not* seem to be the code of interpretation of the corresponding third words in the Tanjore manuscript formulae. The author of the manuscript himself translates his formulae into numbers and this shows that he has a different way of reading the third word in each of his formulae. His methodology however eludes this author's comprehension. However he has been considerate enough to give the equivalent of his formulae in numbers at the end of his exposition of this topic. That is why we are able to reproduce the formulae below in the form of a table analogous to the table of Chapter 3, subject, however to our interpretation of the numbers.

The 27 formulae of the clock of the sky as per the Tanjore manuscript

FORMULA IN SANSKRIT	ENGLISH EQUIVALENT			
	COL. 1	COL. 2	COL. 3	COL. 4
1. Aswini Karke Parvam	Cancer	1	33	37
2. Bharanî Karki Lâbham	Cancer	3	33	85
3. Krittikâ Simhe Kâyah	Leo	1	11	28
4. Rohini Simhe Guruh	Leo	3	15	78
5. Mrigaširah Kanyâ Nîlah	Virgo	0	24	9
6. Ârdrâ Kanyâ Raghuh	Virgo	2	34	61
7. Punarvasu Tulâ Nîtâ	Libra	0	45	18
8. Pushyas Tulâ Gânam	Libra	2	58	71
9. Âslesha Tulâ Gâvah	Libra	3	27	82
10. Maghâ Vrischike Kâmî	Scorpio	1	39	39
11. Pûrvâ Vrischike Bhânu	Scorpio	3	55	94
12. Uttarâ Dhanur Narah	Sagittarius	0	16	6
13. Haste Dhanushi Râmah	Sagittarius	2	37	62
14. Chitrâ Dhanur Bharah	Sagittarius	4	14	101
15. Svâti Makare Kâyah	Capricornus	1	49	43
16. Višakâ Makare Vargam	Capricornus	4	24	105
17. Anuradha Kumbhe Kâmyah	Aquarius	1	6	26
18. Jyeshtâ Kumbhe Râgyâh	Aquarius	2	0	48
19. Mûlam Kumbhe Bhogi	Aquarius	4	22	104
20. Pûrvashadâ Meene Râjyam	Pisces	2	4	49

21.	Uttarashadâm Meene Lolah	Pisces	3	20	80
22.	Šronâ Meshe Kâmah	Aries	1	37	38
23.	Šravishtâ Meshe Gârgih	Aries	3	21	80
24.	Šatabishag Vrishabhe Bhârah	Taurus	4	16	102
25.	Proshtavân Mithune Nabah	Gemini	0	29	11
26.	Ahirbudhnyo Mithune Râkâ	Gemini	2	10	52
27.	Revatî Karke Nâkah	Cancer	0	18	7

Analogous to the table of Chapter 3 consisting of the 27 Sanskrit formulae Col.1 gives the Zodiacal Sign which is rising at the time of observation. Col.2 and Col.3 give the amount by which the sign has risen. *Here arises the problem of interpretation*. In the preceding table we have taken the numbers in Columns 2 and 3 to represent the number of *Ghatikâs* and the number of *Vinâdis* (one-sixtieth of a *Ghatikâ*). This interpretation is only an inference. The manuscript does not seem to say so in the relevant pages. However, this seems to be the most plausible inference. *What is Tulaja-Râja's own interpretation is the million-dollar question*. This is in fact the dark area in understanding the manuscript. At this point it might be worthwhile to look into what precedes Tulaja-raja's numerical answers in his own manuscript. This is how he begins the topic on Folio 98 (of the later version): (In order to get at the time of night from the positions in midsky of the stars we have to calculate the ascending lagna for each position of the midsky).

> *Kha-madhya-sthita-nakshatrât râtri-gata-kâla-jnânârtham kha-madhya-lagna-sâdhanam kartavyam. Tadartham nakshatra-dina-ardhâni kha-madhya-lagnâni ca ..... atha likhyate.*

And he gives the following tabular presentation of his calculation. (Folio 99 & 100). Each column is headed by the first syllable of the nakshatra, (*ašwini*, *bharani*, etc.) relating to which calculations are given below.

a	*bha*	*kr̲*	*ro*	*mr̲*	*â*	*pu*	*pu*	*â*	*ma*
15	15	15	15	15	15	15	15	15	15
21	19	45	45	48	50	45	40	38	25
la	*la*	*la*	*la*	*la*	*la*	*la*	*la*	*la*	*la*
3	3	4	4	5	5	6	6	7	7
8	19	6	18	2	15	4	17	20	9
36	21	31	28	22	17	28	44	34	21
18	0	31	25	51	41	1	47	35	12

pû	*u*	*ha*	*ci*	*swâ*	*vi*	*a*	*jye*	*mû*	*pû*
15	15	14	14	14	14	14	14	14	14
14	5	54	46	32	23	17	15	12	10

la	*la*	*la*	*la*	*la*	*la*	*la*	*la*	*la*	*la*
7	8	8	8	9	9	10	10	10	11
22	1	14	23	10	25	7	12	28	14
12	27	16	2	22	7	7	56	10	42
53	34	54	37	37	37	37	14	24	32

u	*a*	*šra*	*dha*	*ša*	*pû*	*u*	*re*
14	14	14	14	14	14	14	14
12	11	16	20	43	48	57	14
la	*la*	*la*	*la*	*la*	*la*	*la*	*la*
11	11	0	0	1	2	2	3
23	19	11	23	27	2	12	1
41	53	21	44	25	45	25	38
7	21	12	3	20	29	35	27

Uccastha-nakshatra-lagne gata-ghaṯikâh. (*On Folios 101 &102*). The elapsed times of night when the nakshatra is on the meridian are given below by mentioning the sign that is rising and the amount by which it has risen — in the style of our formulae of Chapter 3.

a - karka	*bha - karka*	*kṟ - sim*	*ro - sim*	*mṟ - kam*
1	3	1	3	0
33	33	11	15	24
â - kam	*puna - tu*	*pu -tu*	*â - tu*	*ma - vṟ*
2	0	2	3	1
34	45	58	27	39
pû - vṟ	*u - dha*	*ha -dhas*	*ci - dha*	*swâ - ma*
3	0	2	4	1
55	16	37	14	49
vi - ma	*anu -kum*	*jye - kum*	*mû - kum*	*pû - mî*
4	1	2	4	2
24	6	0	22	4
u - mî	*šra - me*	*dha - me*	*ša - vṟ*	*pû-bhâ - mi*
3	1	3	4	0
20	37	21	16	29
u-bhâ - mi	*re - karka*			
2	0			
10	18			

Etad-eva vâkyatayâ likhyate

(And here it is that we find the Sanskrit statements of the formulae as reproduced in column 1 of pages 48 and 49)

The calculations in the manuscript on Folio 99 seem to suggest the following procedure of ours. Under *Aswini* the entries are in this order: 15, 21, la, 3, 8, 36, 18. The first two numbers are mystifying. The remaining entries may be understood as follows. la, 3, 8, 36, 18 could mean that after three signs are over (from the first, namely, *Mesha*), in the fourth, (namely *Karka*) 8° 36' 18" have risen (when Asvini is on the meridian). If we pursue this logic, then in *Karka* (after making the necessary arithmetical conversion of degrees into *ghatikâs*) the time elapsed should be 1 *ghatikâ* and 26 *vinâdis* (assuming that *Karka* has 5 *ghatikâs* of time-space in all, i.e., two hours of time-space). But this answer 1-26 in *Karka* does not agree with the manuscript's answer 1-33 on Folio 101. If we carry on this calculation for every Nakshatra, following the numbers on Folios 99 and 100, very few of our answers thus derived agree with those on Folios 101 and 102, though we are almost near (!) the answers of the manuscript. This is the crux of the problem in understanding the manuscript.

Now for the comparison with the 27 Sanskrit formulae as per tradition of this author, given in Chapter 3. The differences are only in the *Vinâdi* portion of the formulae. It appears there have been different traditions due to several reasons. One possibility is that the different formulae belong to different epochs. Another possibilty is that the different formulae have been generated for different locations e.g. Tanjore, Ujjain etc. This seems to be a right conjecture because of the statement '*Kshepâscha Tanjâpuryâm*' appearing on Folio number 55 of the Tanjore Manuscript (first version). Another possibility is that the formulae handed down by tradition may have been corrected periodically by actual observation and the different formulae that we have before us may actually belong to different periods where they were in vogue.

But the beauty of all this is the uniformity in the patterns of the formulae both in the Sanskrit and the Tamil tradition as well as the near-accuracy with which they match the actual answers that one can get from a star map. We shall take up the comparative accuracies of the different formulae in the succeeding Chapter.

Chapter 6

A Critical Analysis

In the previous three chapters we have discussed four sets of formulae: The 27 Sanskrit formulae that came from the author's ancestral tradition (we shall christen this SF for the purpose of this chapter), the 27 Tamil formulae (to be christened TF), the Magic Tamil formulae (MTF), and the Tanjore Manuscript formulae (TMF). It is now time to critically examine the four formulae, compare and analyse them from the point of view of their accuracy. What is it that we are comparing? — obviously, their ability to correctly predict the time of night, without reference to a watch, but only with the help of an observation of one of the 27 asterisms near or on the meridian and with the memory of some basic facts about the calendar and the zodiac. The correctness of the prediction can be certainly checked against one's watch provided one knows how to correct it for local time but this would need an actual field testing, where one goes out during night, uses one of these four formulae and arrives at the time of night by the use of that formula. Obviously this field work, though interesting, is bound to be time-consuming and will also consume several man-hours. A better alternative would be to refer to a good star atlas and make the necessary comparisons right at the desk.A still better alternative would be to make the simulations in a planetarium and verify the answers. Essentially this is what has been attempted in the following pages. However, it must be mentioned here that many of these formulae have been used by the author over a lifetime of actual observations out in the field, most often successfully.

The Atlas that has been used to make the desk calculations is 'A Field Guide to the Stars and Planets' of the Peterson Field Guide Series, published by Houghton Mifflin Company, Boston, MA, U.S.A., 1983. The location of the 27 asterisms has been taken to be as described by us in the 2nd Chapter of this book and their times of upper transit has been approximated from the sky maps given in the Field Guide, between pages 36 to 108. Whenever an asterism from our list is just a single star like, say *Rohini* (= Aldebaran) the reference to the sky map yields a fairly accurate time of transit. On the other hand when our asterism is itself a group of stars spread over a longitudinal span over the ecliptic, like, say, *Revati* or *Âsleshâ*, our reading of the time of upper transit from the sky map can at best be only an approximation, sometimes depending on the subjective reading of what constitutes the upper transit of that asterism. The following three pages give, in three tables, the approximate dates of meridian transit of the 27 asterisms at stated hours of the night in hourly intervals, as culled out from these star maps.

APPROXIMATE DATES OF MERIDIAN TRANSIT OF THE 27 ASTERISMS AT STATED HOURS OF THE NIGHT

STAR NO.	1	2	3	4	5	6	7	8	9
HOURS ↓									
18	30-Jan	8-Feb	18-Feb	1-Mar	6-Mar	12-Mar	1-Apr	2-May	6-May
19	15-Jan	24-Jan	3-Feb	14-Feb	19-Feb	25-Feb	17-Mar	17-Apr	21-Apr
20	31-Dec	9-Jan	19-Jan	30-Jan	4-Feb	10-Feb	2-Mar	2-Apr	6-Apr
21	16-Dec	25-Dec	4-Jan	15-Jan	20-Jan	26-Jan	15-Feb	16-Mar	22-Mar
22	1-Dec	10-Dec	20-Dec	31-Dec	5-Jan	11-Jan	31-Jan	1-Mar	7-Mar
23	16-Nov	25-Nov	5-Dec	16-Dec	21-Dec	27-Dec	16-Jan	16-Feb	20-Feb
24	1-Nov	10-Nov	20-Nov	1-Dec	6-Dec	12-Dec	1-Jan	1-Feb	5-Feb
1	15-Oct	24-Oct	3-Nov	14-Nov	19-Nov	25-Nov	15-Dec	15-Jan	19-Jan
2	30-Sep	9-Oct	19-Oct	30-Oct	4-Nov	10-Nov	30-Nov	31-Dec	4-Jan
3	15-Sep	24-Sep	4-Oct	15-Oct	20-Oct	26-Oct	15-Nov	16-Dec	20-Dec
4	31-Aug	9-Sep	19-Sep	30-Sep	5-Oct	11-Oct	31-Oct	1-Dec	5-Dec
5	16-Aug	25-Aug	5-Sep	16-Sep	27-Sep	27-Sep	17-Oct	17-Nov	21-Nov
6	1-Aug	10-Aug	20-Aug	31-Aug	11-Sep	11-Sep	1-Oct	1-Nov	5-Nov

Interpretation

For example star no.4 which is Rohinî will transit the meridian on Dec 31 at 22 hours , on Dec 16 at 23 hours, on Dec 1 at 24 hours and so on . On any intervening date, the transit times may be calculated as per the thumb rule in the Introductory Chapter.

Legend for Star nos.

1. Aświni
2. Bharanî
3. Krittikâ
4. Rohinî
5. Mrigaśirâh
6. Ârdrâ
7. Punarvasu
8. Pushya
9. Âśleshâ

APPROXIMATE DATES OF MERIDIAN TRANSIT OF THE 27 ASTERISMS AT STATED HOURS OF THE NIGHT (contd.)

STAR NO.	10	11	12	13	14	15	16	17	18
HOURS ↓									
18	1-Jun	6-Jun	13-Jun	30-Jun	7-Jul	14-Jul	30-Jul	13-Aug	30-Aug
19	17-May	22-May	29-May	15-Jun	22-Jun	29-Jun	15-Jul	29-Jul	15-Aug
20	2-May	7-May	14-May	31-May	7-Jun	14-Jun	30-Jun	14-Jul	31-Jul
21	17-Apr	22-Apr	29-Apr	16-May	23-May	30-May	15-Jun	29-Jun	16-Jul
22	2-Apr	7-Apr	14-Apr	1-May	8-May	15-May	31-May	14-Jun	1-Jul
23	18-Mar	23-Mar	30-Mar	16-Apr	23-Apr	30-Apr	16-May	31-May	16-Jun
24	3-Mar	8-Apr	15-Mar	1-Apr	8-Apr	15-Apr	1-May	15-May	1-Jun
1	14-Feb	19-Feb	26-Feb	15-Mar	22-Mar	29-Mar	14-Apr	28-Apr	15-May
2	30-Jan	4-Feb	11-Feb	28-Feb	7-Mar	14-Mar	30-Mar	13-Apr	30-Apr
3	15-Jan	20-Jan	27-Jan	13-Feb	19-Feb	26-Feb	14-Mar	28-Mar	14-Apr
4	31-Dec	5-Jan	12-Jan	29-Jan	4-Feb	11-Feb	27-Feb	13-Mar	30-Mar
5	17-Dec	22-Dec	29-Dec	15-Jan	22-Jan	29-Jan	14-Feb	28-Feb	17-Mar
6	1-Dec	6-Dec	13-Dec	30-Dec	6-Jan	13-Jan	29-Jan	12-Feb	1-Mar

Interpretation

For example star no. 16 which is Višaka will transit the meridian on May 31 at 22 hours, on May 16 at 23 hours, on May 1 at 24 hours and so on. On any intervening date , the transit times may be calculated as per the thumb rule in the Introductory Chapter.

Legend for Star nos.

10. Makha
11. Pûrvaphalguni
12. Uttaraphalguni
13. Hasta
14. Chitrâ
15. Swâti
16. Višaka
17. Anurâdhâ
18. Jyeshtâ

APPROXIMATE DATES OF MERIDIAN TRANSIT OF THE 27 ASTERISMS AT STATED HOURS OF THE NIGHT (contd.)

STAR NO. / HOURS ↓	19	20	21	22	23	24	25	26	27
18	13-Sep	29-Sep	5-Oct	18-Oct	30-Oct	22-Nov	5-Dec	30-Dec	18-Jan
19	29-Aug	14-Sep	20-Sep	3-Oct	15-Oct	7-Nov	20-Nov	15-Dec	3-Jan
20	14-Aug	30-Aug	5-Sep	18-Sep	30-Sep	23-Oct	5-Nov	30-Nov	19-Dec
21	30-Jul	15-Aug	21-Aug	3-Sep	15-Sep	8-Oct	21-Oct	15-Nov	4-Dec
22	15-Jul	31-Jul	6-Aug	19-Aug	31-Aug	23-Sep	6-Oct	31-Oct	19-Nov
23	30-Jun	16-Jul	22-Jul	4-Aug	16-Aug	8-Sep	21-Sep	16-Oct	4-Nov
24	15-Jun	1-Jul	7-Jul	20-Jul	1-Aug	24-Aug	6-Sep	1-Oct	20-Oct
1	29-May	14-Jun	20-Jun	3-Jul	15-Jul	7-Aug	20-Aug	4-Sep	23-Sep
2	14-May	30-May	5-Jun	18-Jun	30-Jun	23-Jul	5-Aug	30-Aug	18-Sep
3	28-Apr	14-May	20-May	2-Jun	14-Jun	7-Jul	20-Jul	14-Aug	2-Sep
4	13-Apr	29-Apr	5-May	18-May	30-May	22-Jun	5-Jul	30-Jul	18-Aug
5	31-Mar	16-Apr	22-Apr	5-May	17-May	9-Jun	22-Jun	17-Jul	5-Aug
6	15-Mar	31-Mar	6-Apr	19-May	1-May	24-May	6-Jun	1-Jul	20-Jul

Interpretation

For example star no. 22 which is Šravana will transit the meridian on Aug 19 at 22 hours, on Aug 4 at 23 hours, on July 20 at 24 hours and so on. On any intervening date, the transit times may be calculated as per the thumb rule in the Introductory Chapter.

Legend For Star nos.

19. Mûla
20. Pûrvashada
21. Uttarashada
22. Šravana
23. Šravishta
24. Šatabhishaj
25. Pûrva-Proshtapada
26. Uttara-Proshtapada
27. Revatî

We have used these tables as the base of our comparison for critically analysing the correctnmess or otherwise of the four formulae. In order for the comparisons to be meaningful we shall classify them into *three* methods as detailed in the succeeding pages.

METHOD 1

Compare the answer obtained from each of the formulae with the answer obtained from a reference to the sky maps. For this purpose we have to take certain dates and certain stars or asterisms which we may see transiting the meridian on the night of that date. The various examples we have been considering in the previous chapters come in handy for this purpose. In addition we have taken arbitrarily some more dates and some stars also. All these are tabulated in Table 1 below. Columns 3 to 8 give the time (according to the respective formulae) from 00: 00 to 23: 59 hours. Together all these total upto 36 situations, in each of which we make the comparisons with the Sky Map Answers (christened SMA hereafter). So Table 1 will show the absolute values of SMA - SF, SMA - TF, SMA - MTF, SMA - TMF for each of these 36 cases. Ignore the columns involving the notation MCA for the present. This will be clear when we go to Method 2.

COMPARISON TABLE 1

1 DATE	2 STARS	3 SF	4 TF	5 MTF	6 TMF	7 MCA	8 SMA	9 |SMA -SF|	10 |SMA -TF|	11 |SMA -MTF|	12 |SMA- -TMF|	13 |SMA- -MCA|
1-Jan	PUNARVAS	1:09	1:12	0:06:	1:18	1:20	0:30:	0:39	0:42	0:24	0:48	0:50
5-Jan	MRIGASIRA	22:51	22:54	22:18	22:57	22:56	23:00	0:09	0:06	0:42	0:03	0:04
7-Jan	KRITTIKA	21:07	21:04	20:30	21:08	21:20	20:50	0:17	0:14	0:20	0:18	0:30
12-Jan	ARDRA	23:02	22:56	23:48	23:21	23:04:	22:52	0:10	0:04	0:56	0:29	0:12
30-Jan	MAKHA	1:45:	1:48:	0:48:	1:52	1:36	1:22	0:23	0:26	0:34	0:30	0:14
15-Feb	CHITRA	4:14	4:08	2:24	3:50	4:12	3:45	0:29	0:23	1:21	0:05	0:27
19-Feb	CHITRA	4:06	4:00	2:24	3:42	3:56	3:30	0:36	0:30	1:06	0:12	0:26
27-Feb	VISAKA	5:00	5:00	4:12	5:09	4:56	5:12	0:12	0:12	1:00	0:03	0:16
1-Mar	PUSHYA	22:07	22:04	21:00	22:15	22:12	22:00	0:07	0:04	1:00	0:15	0:12
16-Mar	PUSHYA	21:07	21:04	19:00	21:15	21:12	20:56	0:11	0:08	1:56	0:19	0:16
29-Mar	SWATI	1:59:	1:56:	1:18	2:03	2:00	1:32	0:27	0:24	0:14	0:31	0:28
30-Mar	UTTIRAM	23:27	23:24	22:36	23:18	23:28	22:48	0:39	0:36	0:12	0:30	0:40
8-Apr	JYESHTA	3:17	3:08	4:00	3:20	3:20	3:28:	0:11	0:20	0:32	0:08	0:08
22-Apr	UTTIRASHA	5:13	5:16	4:24	5:00	5:04	5:36	0:23	0:20	1:12	0:36	0:32
1-May	HASTAM	22:21	22:12	21:24	22:02	21:40	22:00	0:21	0:12	0:36	0:02	0:20
15-May	ANUSHAM	0:12	0:12	23:00	0:26	0:20	0:20:	0:08	0:08	1:20	0:06	0:00
18-May	SWATI	22:39	22:36	21:12	22:43	22:40	22:40	0:01	0:04	1:28	0:03	0:00
22-May	PURVA-PHA	19:30	19:12	17:00	19:34	19:18	19:18	0:12	0:06	2:18	0:16	0:00
14-Jun	PURVASHA	0:39	0:24	23:42	1:01	0:28:	1:00	0:21	0:36	1:18	0:01	0:32
7-Jul	SATABHISH	3:47	3:44	3:24	4:26	3:40	3:32	0:15	0:12	0:08	0:54	0:08

16-Jul	JYESHTA	20:53	20:44	19:54	20:56	20:44	20:56	0:03	0:12	1:02	0:00	0:12
24-Jul	U.PROSHTA	5:16	5:16	3:06	4:32	5:12	5:24	0:08	0:08	2:18	0:52	0:12
1-Aug	SRAVISHTA	1:24	1:34	0:24:	0:42	1:04	0:00	1:24	1:34	0:24	0:42	1:04
14-Aug	MOOLAM	20:55	20:58	18:42	21:06	19:36	20:24	0:31	0:34	1:42	0:42	0:48
3-Sep	SRAVANA	22:02	22:08	21:24	21:34	21:48	21:08	0:54	1:00	0:16	0:26	0:40
16-Sep	ROHINI	5:30	5:30	2:22	5:30	5:44	5:04	0:26	0:26	2:42	0:26	0:40
18-Sep	REVATI	2:43:	2:22	0:54:	2:22:	2:36	2:20	0:23	0:02	1:26	0:02	0:16
30-Sep	ASWINI	2:11	2:08	0:48:	1:57	1:56	2:24	0:13	0:16	1:36	0:27	0:28
21-Oct	P..PROSHTA	22:26	22:32	20:12	22:07	22:12	20:44	1:42	1:48	0:32	1:23	1:28
24-Oct	MAGHA	8:17	8:20	7:00	8:23	8:12	8:00	0:17	0:20	1:00	0:23	0:12
26-Oct	ARDRA	4:18	4:12	3:18	4:37	4:16	4:00	0:18	0:12	0:42	0:37	0:16
20-Nov	KRITTIKA	0:33:	0:30:	22:36	0:34:	23:24	23:44:	0:49	0:46	1:08	0:50	0:20
21-Nov	ASLESHA	5:37	5:34	4:00	5:14	5:36	4:36	1:01	0:58	0:36	0:38	1:00
30-Nov	U.PROSHTA	20:52	20:52	19:00	20:08	20:36	20:04	0:48	0:48	1:04	0:04	0:32
11-Dec	ROHINI	23:46	23:46	21:24	23:46	0:00	23:16	0:30	0:30	1:52	0:30	0:44
25-Dec	BHARANI	21:39	21:18	19:36	21:01	21:36	20:40	0:59	0:38	1:04	0:21	0:56

METHOD 2

This involves what is called the **Master Chart**. (See Pull-out after page 62). The Chart purports to do the following. It is actually the ecliptic in the form of a circle. The 27 Nakshatras occupy equal portions of the ecliptic, namely 13° 60′. Two and one-fourth Nakshatras make one sign of the Zodiac and these are also marked and named. The entire circle is divided into 365 divisions each corresponding to one day of the year and approximately one degree of circular arc. This much is the base on which the remaining information is superposed. These will change from era to era. For instance, the First Point of Aries, (the Vernal Equinox) which is a relatively fixed point of the ecliptic as far as the movement of the Sun is concerned, is also marked. The Sun is at this point on March 21. The dates of the calendar are now marked around the circle. (But several years from now, the Sun will be at this point on March 20, due to precession of the equinoxes — the situation actually changes by one full day every 72 years). So right now, starting from March 21, the remaining days of the year are marked. The position March 21 also corresponds to the point in the Nakshatra, which is the end of *Purva Proshtapadâ* and the beginning of *Uttara-Proshtapadâ*, because that is where the vernal equinox is now with respect to the distant stars. In other words the Sun on March 21 is at the point about 3 degrees into Meena Râsi (Pisces sign of the Zodiac). Summarising all this we may say that the chart converts the position of the Sun round the year on the ecliptic into a calendar date x and these are what are shown in the chart.

In the same way we convert the second parameter (say, of the formula SF) also as a calendar date. The second parameter is just the eastern horizon point indicated by the Nakshatra seen on the meridian. For instance, when Punarvasu is on the meridian, the eastern horizon point (on the zodiac or the ecliptic) is Neelâ times from the beginning of Tulâ, i.e. 9 minutes of (daily motion) time into Libra. This point is already available on the chart since each sign of the zodiac is marked and is also calibrated in terms of days of the calendar. Note that that 4 minutes of (daily motion) time is equivalent to one day in this chart. So the point P indicated by the formula is just a calendar date *y* on the circle. These 27 calendar dates are indicated on the chart by an extended line (corresponding to the appropriate date) towards the centre across which the name of a Nakshatra is written. This Nakshatra is the one on the meridian from which we got this indication through the formula. For instance, at the *Neelâ* point in *Tulâ* — which corresponds to the calendar date of October 22 — we extend the radius inward and write on it, Punarvasu. This is done for each of the 27 formulae and so we have 27 inwardly extended radii on which the names of the 27 Nakshatras are written.

The only thing to be done in order to use this chart is now to identify the two parameters — namely, the date on which the observation is made and the Nakshatra that is seen on the meridian — in terms of two calendar dates on this chart and calculate the difference x-y as a certain number of days. This number is to be considered as so many degrees and a division by 15 gives the number of hours and minutes before sunrise. Thus we know the time of night.

For instance, suppose we see *Anurâdhâ* Nakshatra on May 15 transiting the meridian. The Anurâdhâ marking on the chart corresponds to February 19. The number of days from February 19 to May 15 is 85. Dividing by 15 we get 5 hours and 40 minutes. The time of observation is therefore 12-20 A.M.

Thus this is another way of looking at the formulae SF, TF and TMF. The answers obtained thus are called MCA — Master Chart Answers — and are incorporated in Table 2 below and already in Table 1. In essence these are not different from the corresponding formulae from which they are obtained as is shown by the entries SMA - MCA in Table 1 and MCA - SF in Table 2.

COMPARISON TABLE 2

a DATE	b STARS	c SF	d TF	e MTF	f TMF	g MCA-	h |MCA- SF|	i |MCA- TF|	j MCA- MTF|	k |MCA -TMF|
1-Jan	Punarvasu	1:09	1:12	0:06:	1:18	1:20	0:11	0:08	1:14	0:02
5-Jan	Mrigaširah	22:51	22:54	22:18	22:57	22:56	0:05	0:02	0:38	0:01
7-Jan	Krittikâ	21:07	21:04	20:30	21:08	21:20	0:13	0:16	0:50	0:12
12-Jan	Ârdrâ	23:02	22:56	23:48	23:21	23:04	0:02	0:08	0:44	0:17
30-Jan	Maghâ	1:45:	1:48:	0:48:	1:52	1:36	0:09	0:12	0:48	0:16
15-Feb	Chitrâ	4:14	4:08	2:24	3:50	4:12	0:02	0:04	1:48	0:22
19-Feb	Chitrâ	4:06	4:00	2:24	3:42	3:56	0:10	0:04	1:32	0:14
27-Feb	Višâkhâ	5:00	5:00	4:12	5:09	4:56	0:04	0:04	0:44	0:13
1-Mar	Pushya	22:02	22:04	21:00	22:15	22:12	0:10	0:08	1:12	1:15
16-Mar	Pushya	21:07	21:04	19:00	21:15	21:12	0:05	0:08	2:12	0:03
29-Mar	Swâti	1:59:	1:56:	1:18	2:03	2:00	0:01	0:04	0:42	0:03
30-Mar	Uttiram	23:27	23:24	22:36	23:18	23:28	0:01	0:04	0:52	0:10
8-Apr	Jyeshtâ	3:17	3:08	4:00	3:20	3:20	0:03	0:12	0:40	0:00
22-Apr	Uttirâshâdh	5:13	5:16	4:24	5:00	5:04	0:09	0:12	0:40	0:04
1-May	Hastam	22:21	22:12	21:24	22:02	21:40	0:41	0:32	0:16	0:22
15-May	anurâdhâ	0:12	0:12	23:00	0:26	0:20:	0:08	0:08	1:20	0:06
18-May	Swâti	22:39	22:36	21:12	22:43	22:40	0:01	0:04	1:28	0:03
22-May	Pûrva-phal	19:30	19:12	17:00	19:34	19:18	0:12	0:06	2:18	0:16
14-Jun	Pûrvâshâd	0:39	0:24	23:42	1:01	0:28:	0:11	0:04	0:46	0:33
7-Jul	Šatabhishaj	3:47	3:44	3:24	4:26	3:40	0:07	0:04	0:16	0:46
16-Jul	Jyeshtâ	20:53	20:44	19:54	20:56	20:44	0:09	0.00	0:50	0:12
24-Jul	U.Proshta	5:16	5:16	3:06	4:32	5:12	0:04	0:04	2:06	0:40
1-Aug	Šravishta	1:34	1:34	0:24:	0:42	1:04	0:30	0:30	0:40	0:22
14-Aug	Moolam	20:55	20:58	18:42	21:06	19:36	1:19	1:22	0:54	1:30
3-Sep	Šravana	22:02	22:08	21:24	21:34	21:48	0:14	0:20	0:24	0:14
16-Sep	Rohinl	5:30	5:30	5:30	5:30	5:44	0:14	0:14	0:14	0:14
18-Sep	Revatl	2:43:	2:22	0:54:	2:22:	2:36	0:07	0:14	1:42	0:14
30-Sep	Ašwini	2:11	2:08	0:48:	1:57	1:56	0:15	0:12	1:08	0:01
21-Oct	P.Proshtapa	22:26	22:32	20:12	22:07	22:12	0:14	0:20	2:00	0:05
24-Oct	Maghâ	8:17	8:20	7:00	8:23	8:12	0:05	0:08	1:12	0:11
26-Oct	Ârdrâ	4:18	4:12	3:18	4:37:	4:16	0:02	0:04	0:58	0:21
20-Nov	Krittikâ	0:33:	0:30:	22:36	0:34:	23:24	1:09	1:06	0:48	1:10
21-Nov	Âšleshâ	5:37	5:34	4:00	5:14	5:36	0:01	0.02	1:36	0:22
30-Nov	U.Proshtapa	20:52	20:52	19:00	20:08	20:36	0:16	0:16	1:36	0:28
11-Dec	Rohini	23:46	23:46	21:24	23:46	0:00	0:14	0:14	2:36	0:14
25-Dec	Bharanî	21:39	21:18	19:36	21:01	21:36	0:03	0:18	2:00	0:35

METHOD 3

This is suggested by the Master Chart technique. Since the Master Chart converts the formulae (SF, TF, TMF) into calendar dates, they are amenable to a verification by simulation inside a planetarium. This was done in the Nehru Planetarium, New Delhi. Actually the simulation was done for the Tanjore latitude. As was mentioned in Chapter 5, the Tanjore latitude must have had something to do with these formulae, at least one of them, the TMF. Instead of looking at the Master Chart for the exact date which is indicated by the formula, we set the equipment in the sky-theatre of the planetarium for the particular star to be on the meridian on the particular day and note what part of the ecliptic is rising at that time. Since the ecliptic is graduated (on the simulated sky) in the same manner as our Master Chart is, we get the correct date that should have been indicated by the Nakshatra on the meridian. Comparing this with the MCA date we have a comparison of the formula with actuality, in some sense. This comparison, is shown both for SF and TMF in Table 3. The fact of the matter is, this comparison is not a comparison of the answers derived from the different formulae, but a comparison between the formulae themselves.

COMPARISON TABLE 3

Stars	Date as per traditional SF Col 2	Date as per simulation in the Planetarium Col 3	Date as per TMF Col 4	Difference Col 2 - Col 3	Difference Col 4 - Col 3
Aświni	31-Jul	26-Jul	28-Jul	5	2
Bharani	21-Aug	5-Aug	9-Aug	16	4
Krittikâ	30-Aug	19-Aug	31-Aug	11	12
Rohini	12-Sep	1-Sep	12-Sep	11	11
Mrigasirah	21-Sep	18-Sep	22-Sep	3	4
Ârdrâ	30-Sep	22-Sep	5-Oct	8	13
Punarvasu	22-Oct	17-Oct	24-Oct	5	7
Pushya	3-Nov	1-Nov	7-Nov	2	6
Âsleshâ	15-Nov	12-Nov	9-Nov	3	3
Maghâ	26-Nov	22-Nov	27-Nov	4	5
P.Phalguni	10-Dec	6-Dec	11-Dec	4	5
U.Phalguni	21-Dec	14-Dec	19-Dec	7	5
Hastâ	5-Jan	22-Dec	2-Jan	14	21
Chitrâ	18-Jan	4-Jan	13-Jan	14	9
Swâti	26-Jan	17-Jan	28-Jan	9	11
Višâka	11-Feb	25-Jan	13-Feb	17	19
Anurâdhâ	19-Feb	12-Feb	25-Feb	7	13
Jyeshta	27-Feb	21-Feb	3-Mar	6	11
Moolam	11-Mar	12-Mar	17-Mar	1	5
Pûrvâshadha	22-Mar	21-Mar	29-Mar	1	8

Uttarâshâdha	7-Apr	6-Apr	6-Apr	1	0
Šravana	3-May	23-Apr	26-Apr	10	3
Šravishta	20-May	7-May	7-May	13	0
Shatabishaj	1-Jun	29-May	11-Jun	3	13
P.Proshtapada	26-Jun	11-Jun	20-Jun	15	9
U.Proshtapada	11-Jul	27-Jun	1-Jul	14	4
Revati	29-Jul	12-Jul	20-Jul	17	8

CONCLUSION

We can draw several and varied conclusions from all that has been done so far. First let us put away the MTF (Magic Tamil Formula). This formula is worthwhile only for its algorithmic approach and its capacity to give an instantaneous answer. But by that very fastness it loses the accuracy — as has already been pointed out in that chapter — and so gives only a crude approximation.

Putting this formula aside we note that the other formulae namely, SF, TF, and TMF are applicable even today with a reasonable accuracy. The average accuracy turns out to be 25 minutes or so The precision with which these formulae can be applied even today is astounding. Note that in Table 1 above, Column 9 which shows the absolute values of the difference SMA - SF has a total of 36 entries which distribute as follows: (*)

0 to 15 minutes	:	13 entries
16 to 30 minutes	:	12 entries
31 to 45 minutes	:	4 entries
46 to 60 minutes	:	4 entries
greater than 60 min.	:	3 entries

This averages out to about 25 minutes. The sample of 36 examples we have chosen are supposed to have been randomly chosen but the randomness may not satisfy any mathematical testing criterion. However, we get the general impression that the formulae are accurate but for an average error of, say, 30 minutes, this way or that way. Taking into consideration the cultural requirements of the people who worked with these formulae in ancient times, and taking into account the fact that they did all that without the least external help in the form of any equipment, we cannot but marvel at their achievement. The fact that the formulae are only star-dependent is what makes them applicable even several millenia after they were discovered (or invented?). When it says Krittikâ Simhe Kâyâ, the *simhe-kâyâ* has been arrived at by them just by calculating an arc of 90 degrees behind the location of *Krittikâ* on the ecliptic. The formula *ipso facto* has nothing to do with the position on Earth from which the star is observed on the meridian nor does it matter at what local time you observe it. This is the strength of the formulae.

So much for the positive side. Now let us look at some negatives. There are three cases even in the small sample of 36 situations, which give an error of more than one hour. In one case the error is as much as $1^h\ 42^m$. The other two errors are $1^h\ 24^m$ and $1^h\ 01^m$. These happen in the case of *Uttara Proshtapadâ* on 30th November, *Sravishtâ* on 1st August and *Âsleshâ* on 21st November. The first question that arises in one's mind is whether the formulae relating to these particular Nakshatras are wrong. Recall here the fact that the formulae SF have been handed down by oral tradition. So at this point one compares the corresponding entries for the Tanjore Manuscript Formulae. These are 1h 23^{m}, $00^{h}42^{m}$, and 00^{h} 38^{m} respectively. The TMF seems to be better. The distribution corresponding to (*) in the case of the TMF is the following (**):

0 to 15 minutes	:	13 entries
16 to 30 minutes	:	12 entries
31 to 45 minutes	:	6 entries
46 to 60 minutes	:	4 entries
greater than 60 min	:	1 entry.

But this type of detailed statistical comparison is only fruitless hairsplitting. It is not warranted because our examples are not statistically random. We can only go by the general impression. Both the SF and the TMF (The TF goes along with them) are certainly not precise from any scientific standards But they have been used pretty heavily in ancient times, it appears. A historical novel by Ra. Ki. Rangarajan (in Tamil) uses such ideas at a crucial moment in the development of the story that this author wondered where the author of the novel got the idea about keeping track of time of night by a close look at the stars. It turned out that Rangarajan got the idea from Makaram Rajagopalan who has imbibed these ideas by oral tradition only. There are several examples that can be cited like this. The fact that the Tanjore manuscript by Tulaja Râja has thought fit to elaborate on this in a manuscript on astrology shows the prevalence of such methods in olden times. In spite of the fact therefore that several examples give an error of more than 30 minutes we have to appreciate the talent of the ancients who discovered all these. The errors perhaps are due to one or more of four reasons: (1) Our inaccurate interpretation of what constitutes the named asterism of the ancients; (2) The formulae perhaps have been periodically changed by actual observation and for the purpose of taking account of change in the epoch (due to precession of the equinoxes) and so what we have before us may not be the one applicable now and (3) Different formulae might have been in use in different parts of India and the ones that we lay our hands on may not be the right ones to use in the particular location where we want to use them because of the errors in the assumption regarding times of sunrise and sunset. (4) The corrections indicated by the verse on page 34 have not been understood precisely either in terms of the declinations of the *nakshatras* or otherwise and consequently have not been carefully applied.

Thus we come to the end of our story. As we remarked at the beginning the whole purpose of the book has been to show that the ancients of India have used their observational powers remarkably well to put into a mathematical symbolism the rhythm in the daily procession of the stars (at least the 27 stars of Indian astronomy) and relate it to both the annual and diurnal motion of the Sun, thus arriving at beautiful formulae for finding the time of night by one look at the stars in the meridian. This itself is an intellectual marvel put to the most mundane use. Never perhaps was so much achieved with so little so early in the Ascent of Man.

Appendix I

(Ref. page 4)

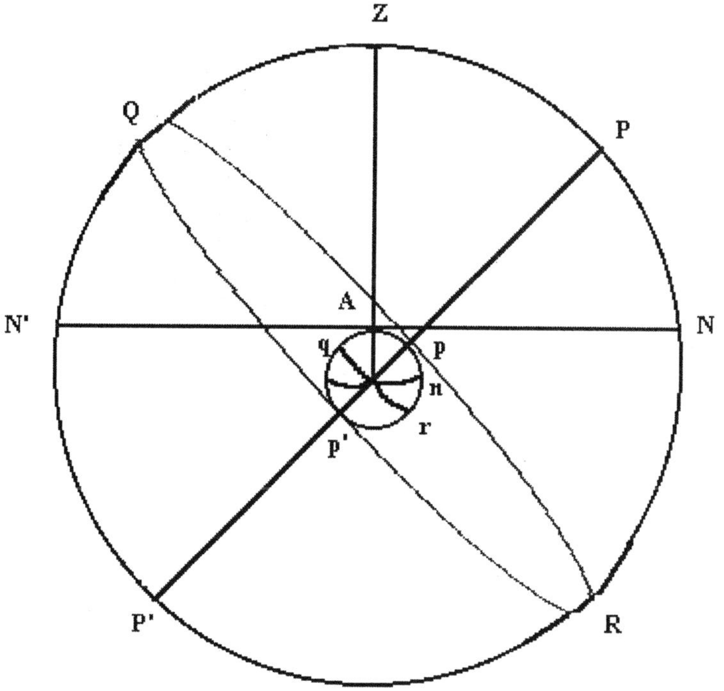

Altitude of Celestial North Pole
=
Geographical Latitude of Plane of observation

The Celestial equator QR is the extension of the equatorial plane (on the earth) in all directions in space. PP′ is the celestial axis around which everything in the sky revolves daily. P is the Celestial North Pole. A is the observer on the Earth. His horizon is NAN′.

The altitude of the Celestial North Pole at any point of observation is the angle between the observer's line of vision to the Celestial pole and the plane of the horizon at the point of observation. So Arc NP is the altitude of the Celestial North pole. Arc qA on Earth is the geographical latitude of A. We have to prove these two are equal when measured by the angles subtended at the centre by these arcs

We have

Arc PN = 90° — Arc ZP (measured by the angle subtended at the centre)

= 90° — Angle AO p

= Angle q O A

AS WINI

Appendix II
(27 pages)

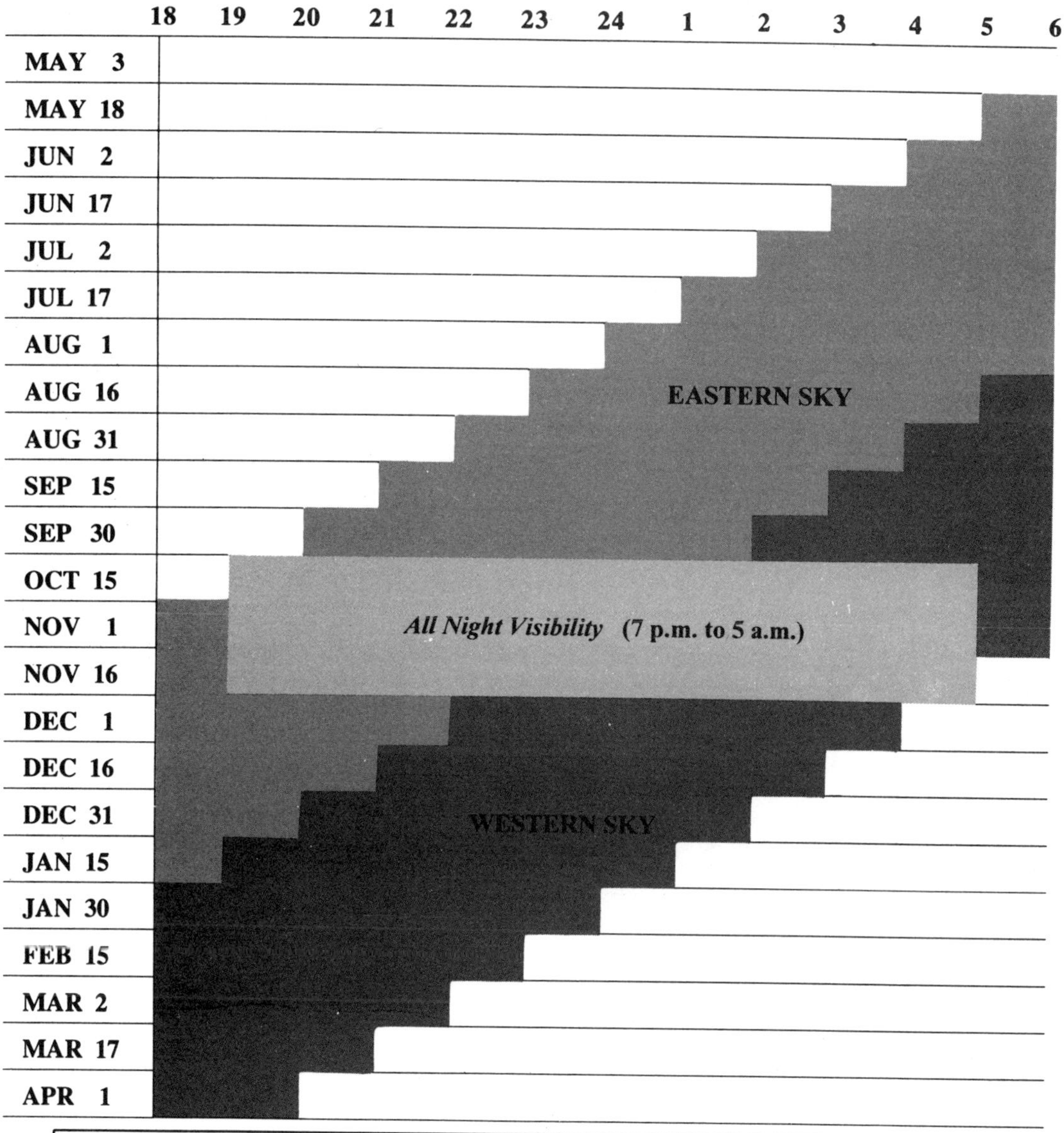

LEGEND (Applicable for all the 27 charts)

Each of these charts gives for the Nakshatra under consideration, the approximate dates of the year and the portion of the sky (Eastern or Western) where one can look for them at the various times of the night — 6 p.m. to 6 a.m. (shown at the top row as 18, 19 etc.) on these dates.

In addition the upper transit time of the Nakshatra can be read from the chart as follows. For each date look along the corresponding horizontal row. The upper transit time is the time corresponding to the vertical line separating the light and dark shades.

For example, on Dec. 16, the upper transit time of Aswini is 9 p.m.

For intermediate dates that are not listed, make an appropriate interpolation.

BHARANI

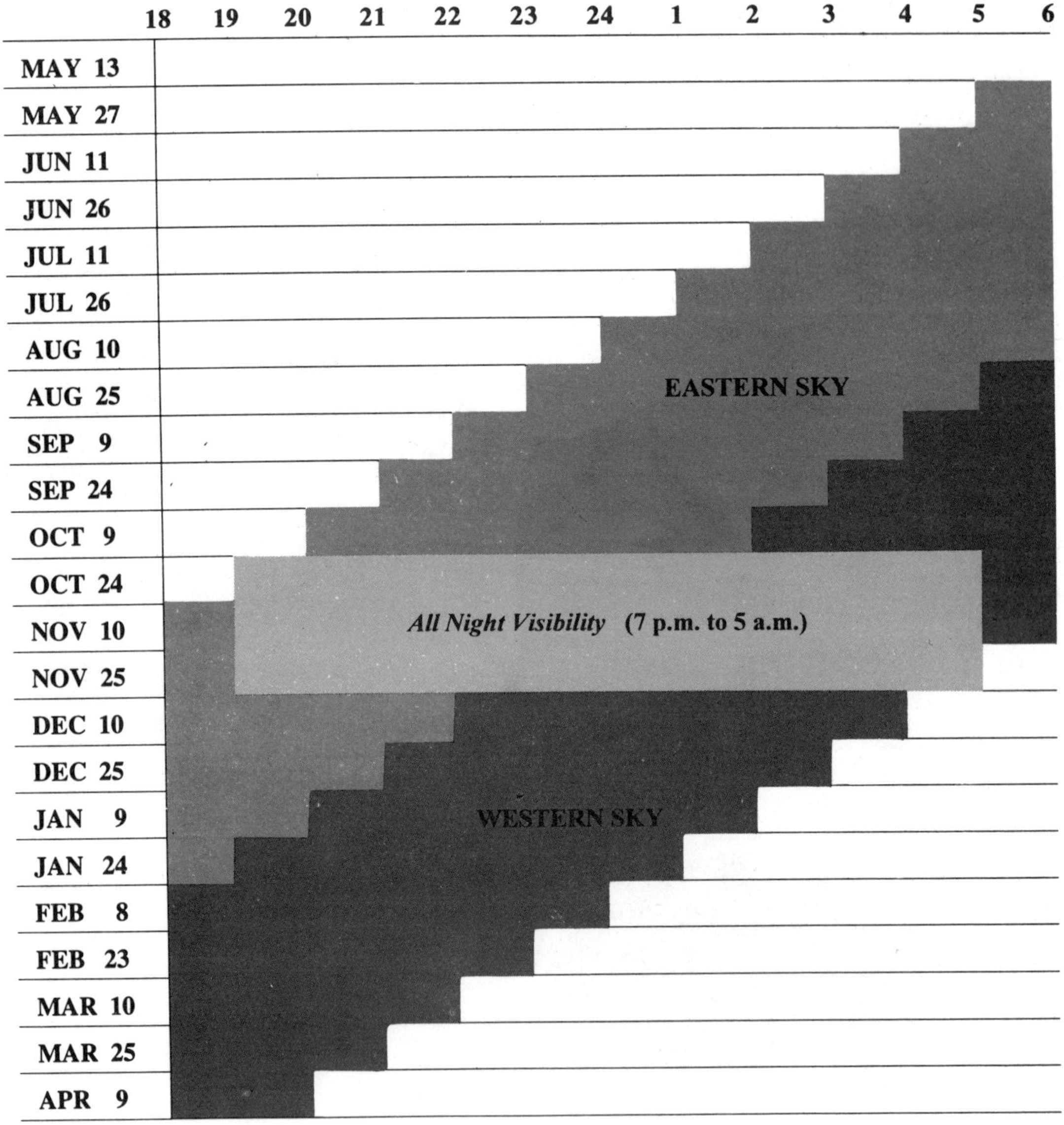
18
19
20
21
22
23
24
1
2
3
4
5
6
MAY 13
MAY 27
JUN 11
JUN 26
JUL 11
JUL 26
AUG 10
AUG 25
SEP 9
SEP 24
OCT 9
OCT 24
NOV 10
NOV 25
DEC 10
DEC 25
JAN 9
JAN 24
FEB 8
FEB 23
MAR 10
MAR 25
APR 9
EASTERN SKY
All Night Visibility (7 p.m. to 5 a.m.)
WESTERN SKY

KRITTIKA

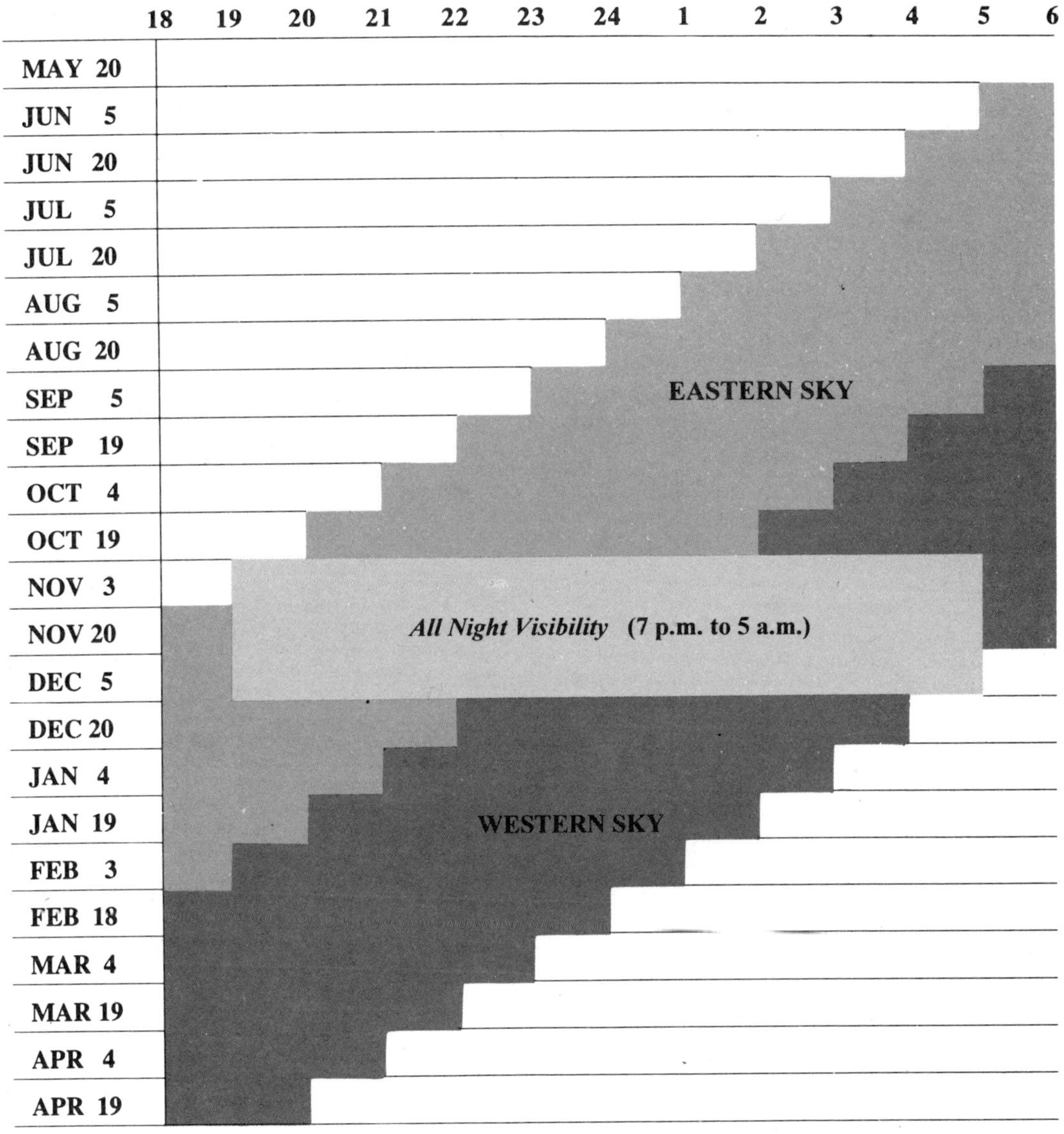
18 19 20 21 22 23 24 1 2 3 4 5 6
MAY 20
JUN 5
JUN 20
JUL 5
JUL 20
AUG 5
AUG 20
SEP 5
SEP 19
OCT 4
OCT 19
NOV 3
NOV 20
DEC 5
DEC 20
JAN 4
JAN 19
FEB 3
FEB 18
MAR 4
MAR 19
APR 4
APR 19
EASTERN SKY
All Night Visibility (7 p.m. to 5 a.m.)
WESTERN SKY

ROHINI

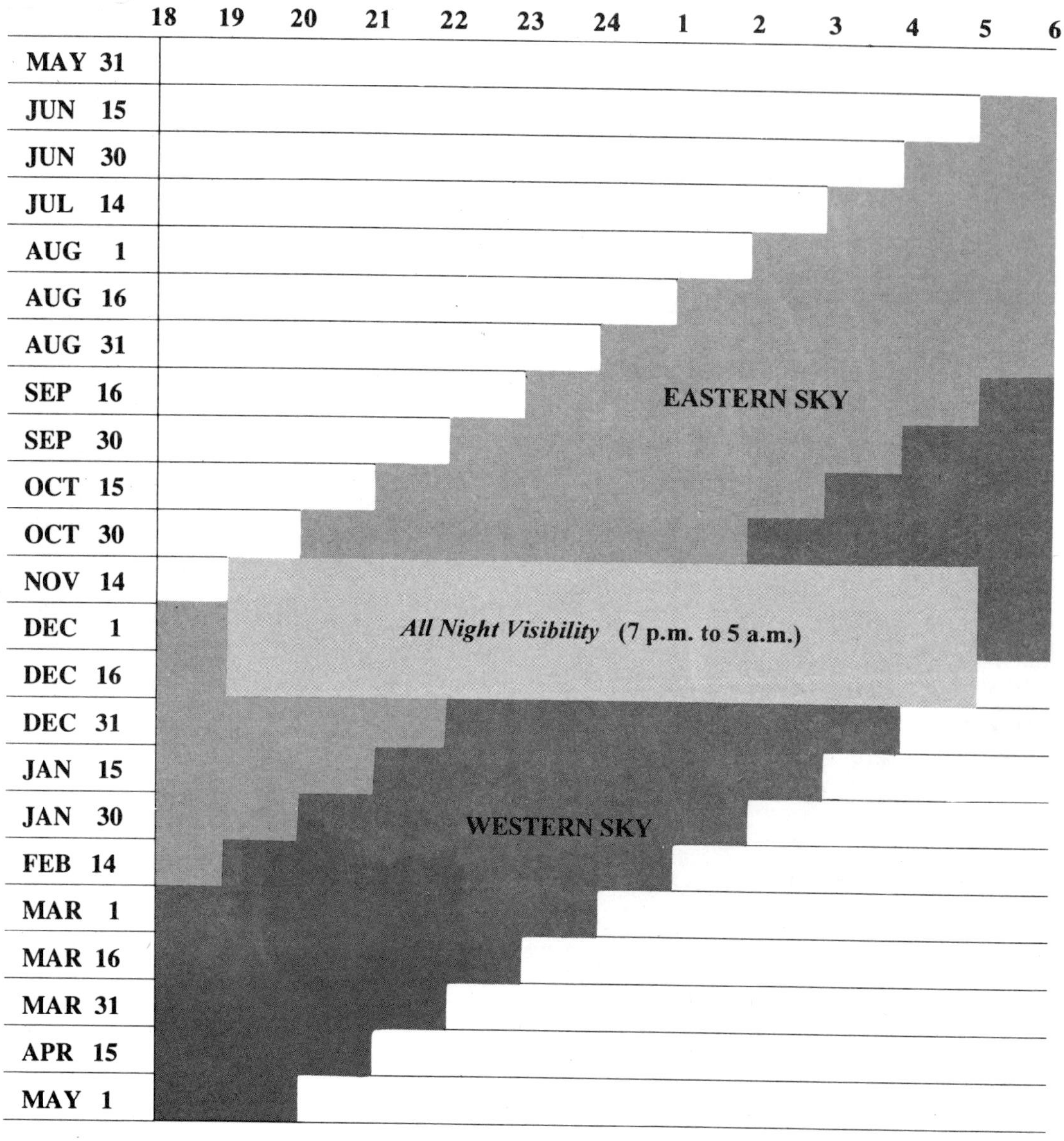
18
19
20
21
22
23
24
1
2
3
4
5
6
MAY 31
JUN 15
JUN 30
JUL 14
AUG 1
AUG 16
AUG 31
SEP 16
SEP 30
OCT 15
OCT 30
NOV 14
DEC 1
DEC 16
DEC 31
JAN 15
JAN 30
FEB 14
MAR 1
MAR 16
MAR 31
APR 15
MAY 1
EASTERN SKY
All Night Visibility (7 p.m. to 5 a.m.)
WESTERN SKY

MRIGASIRAH

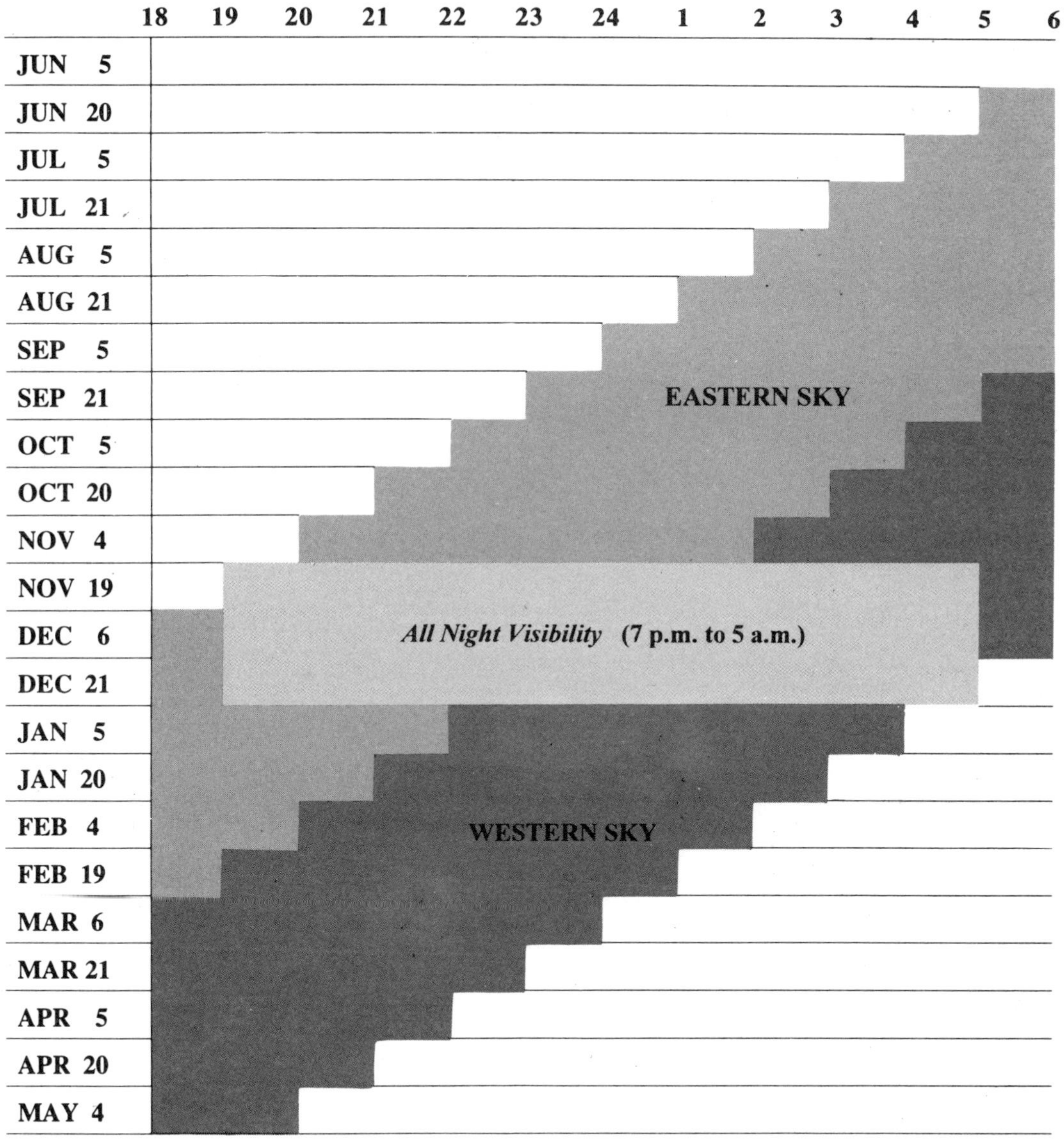

18
19
20
21
22
23
24
1
2
3
4
5
6
JUN 5
JUN 20
JUL 5
JUL 21
AUG 5
AUG 21
SEP 5
SEP 21
OCT 5
OCT 20
NOV 4
NOV 19
DEC 6
DEC 21
JAN 5
JAN 20
FEB 4
FEB 19
MAR 6
MAR 21
APR 5
APR 20
MAY 4
EASTERN SKY
All Night Visibility (7 p.m. to 5 a.m.)
WESTERN SKY

ARDRA

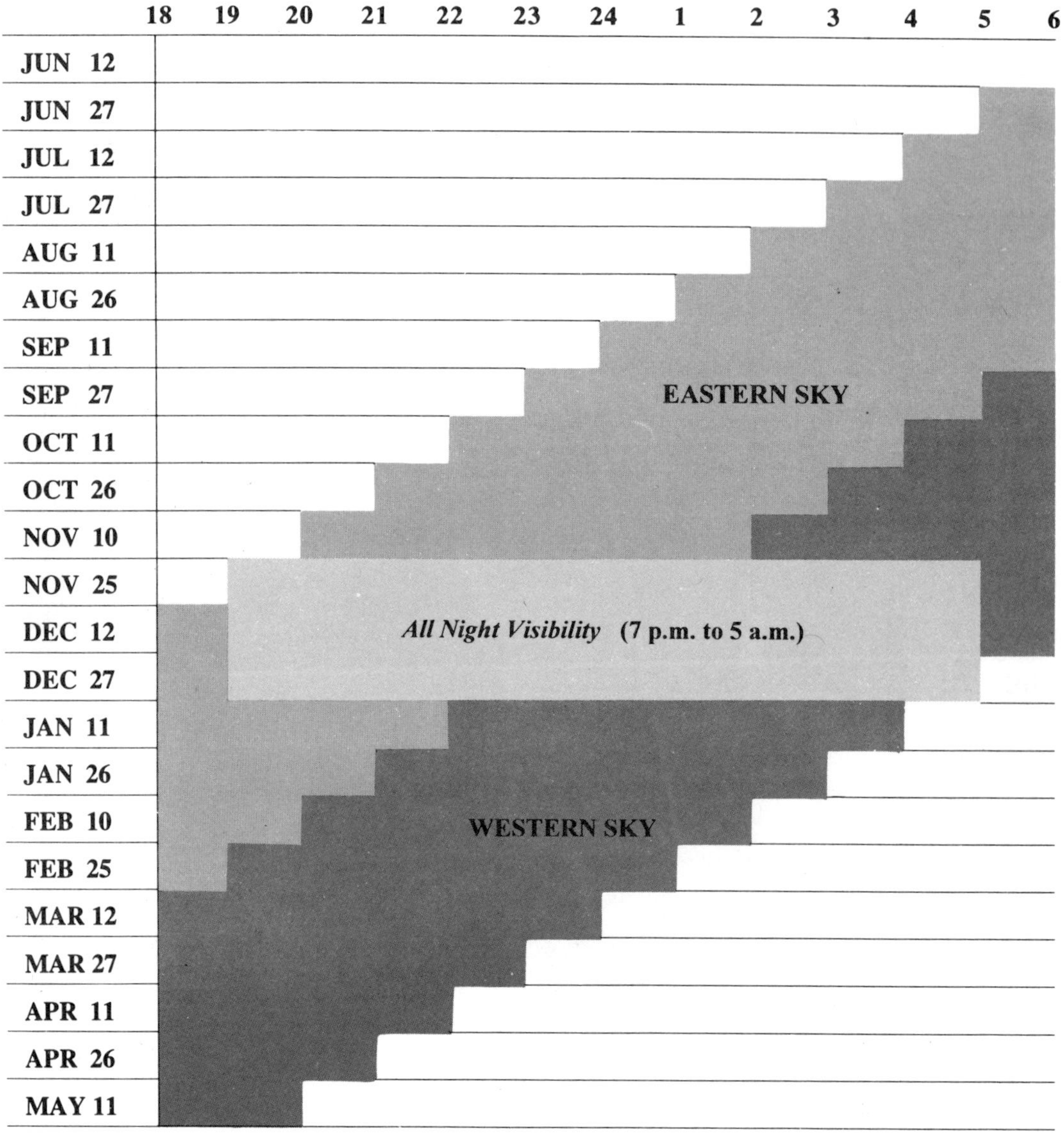
18 19 20 21 22 23 24 1 2 3 4 5 6
JUN 12
JUN 27
JUL 12
JUL 27
AUG 11
AUG 26
SEP 11
SEP 27
OCT 11
OCT 26
NOV 10
NOV 25
DEC 12
DEC 27
JAN 11
JAN 26
FEB 10
FEB 25
MAR 12
MAR 27
APR 11
APR 26
MAY 11
EASTERN SKY
All Night Visibility (7 p.m. to 5 a.m.)
WESTERN SKY

PUNARVASU

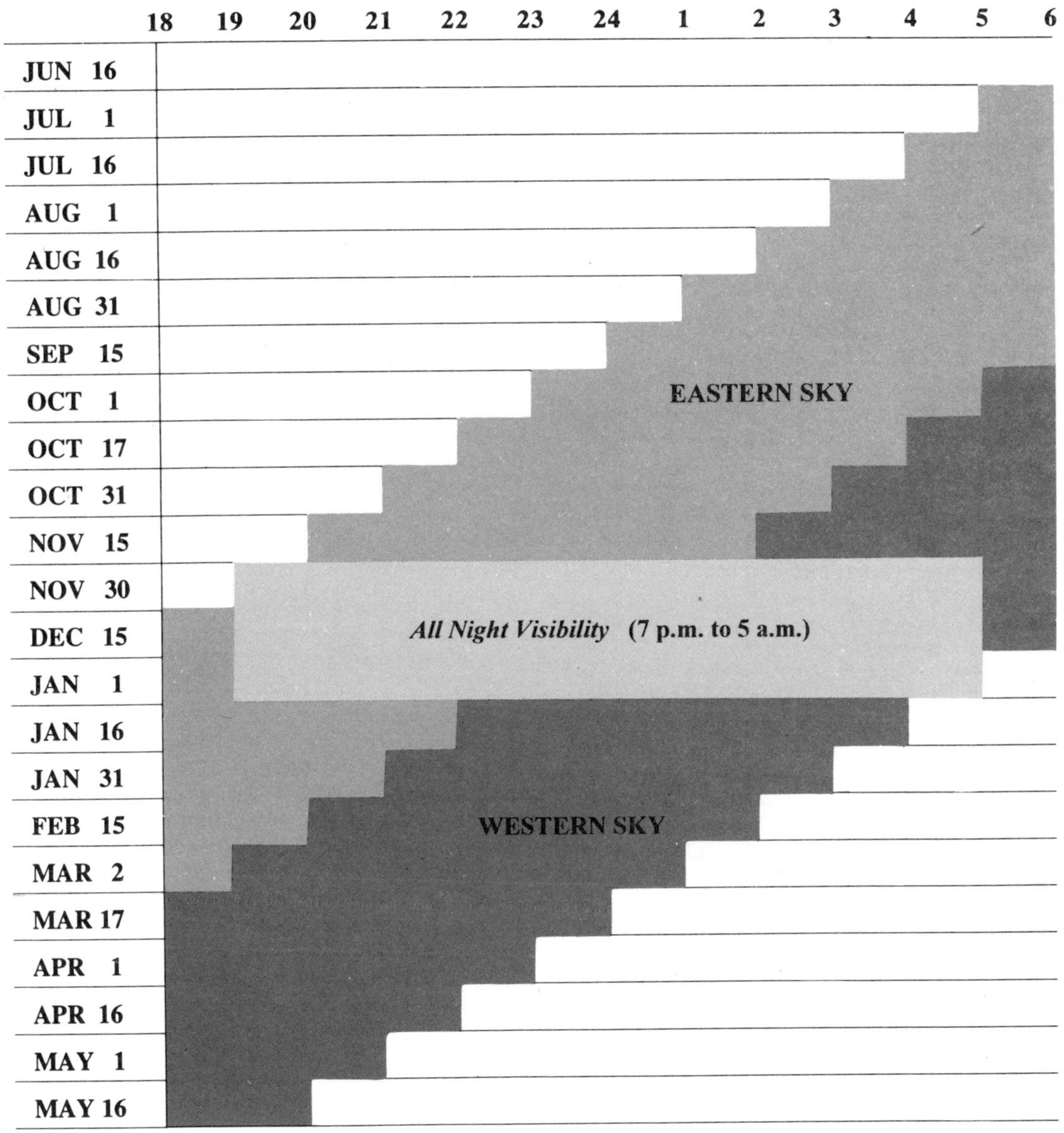
18
19
20
21
22
23
24
1
2
3
4
5
6
JUN 16
JUL 1
JUL 16
AUG 1
AUG 16
AUG 31
SEP 15
OCT 1
OCT 17
OCT 31
NOV 15
NOV 30
DEC 15
JAN 1
JAN 16
JAN 31
FEB 15
MAR 2
MAR 17
APR 1
APR 16
MAY 1
MAY 16
EASTERN SKY
All Night Visibility (7 p.m. to 5 a.m.)
WESTERN SKY

PUSHYA

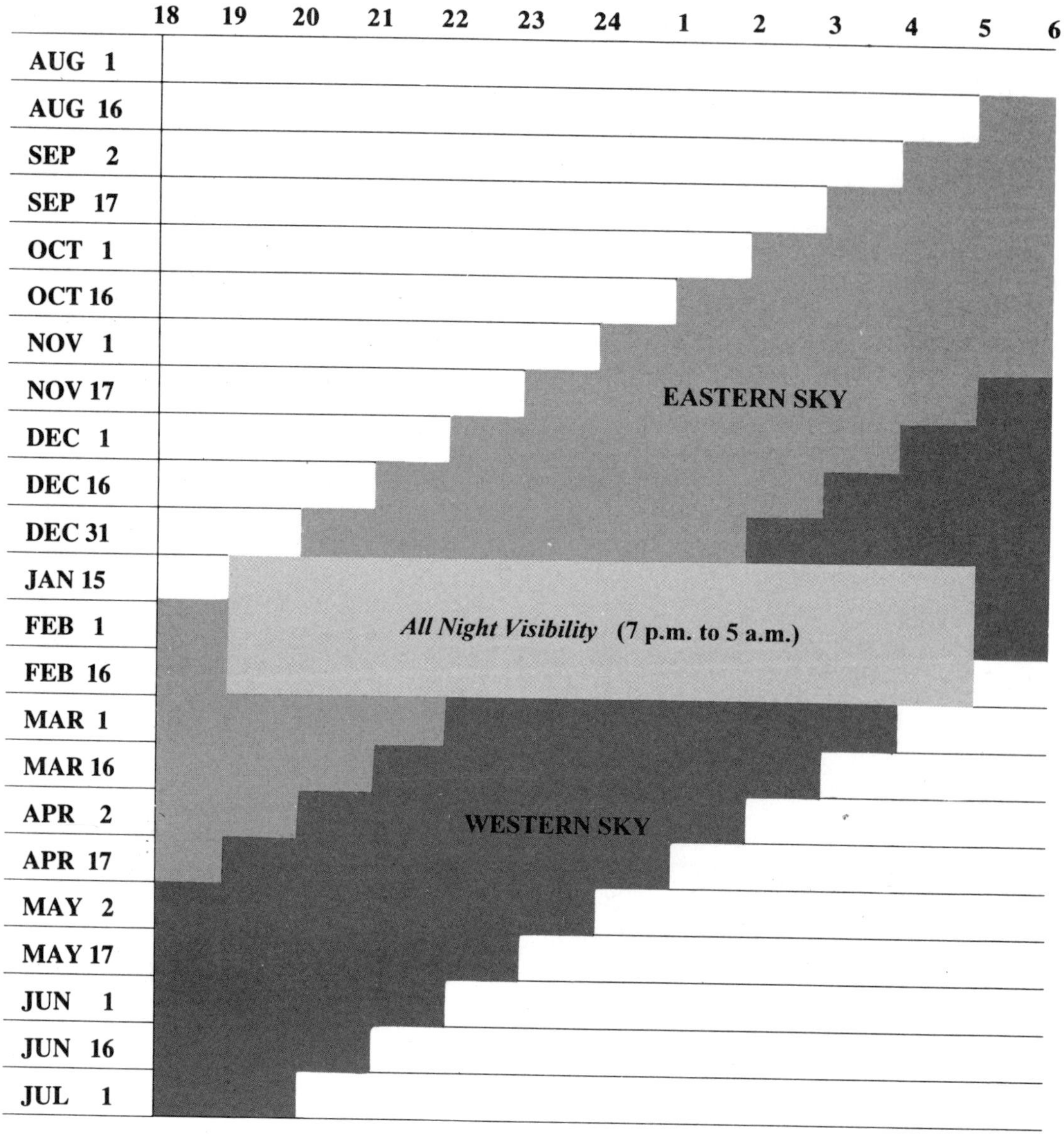
18
19
20
21
22
23
24
1
2
3
4
5
6
AUG 1
AUG 16
SEP 2
SEP 17
OCT 1
OCT 16
NOV 1
NOV 17
DEC 1
DEC 16
DEC 31
JAN 15
FEB 1
FEB 16
MAR 1
MAR 16
APR 2
APR 17
MAY 2
MAY 17
JUN 1
JUN 16
JUL 1
EASTERN SKY
All Night Visibility (7 p.m. to 5 a.m.)
WESTERN SKY

ASLESHA

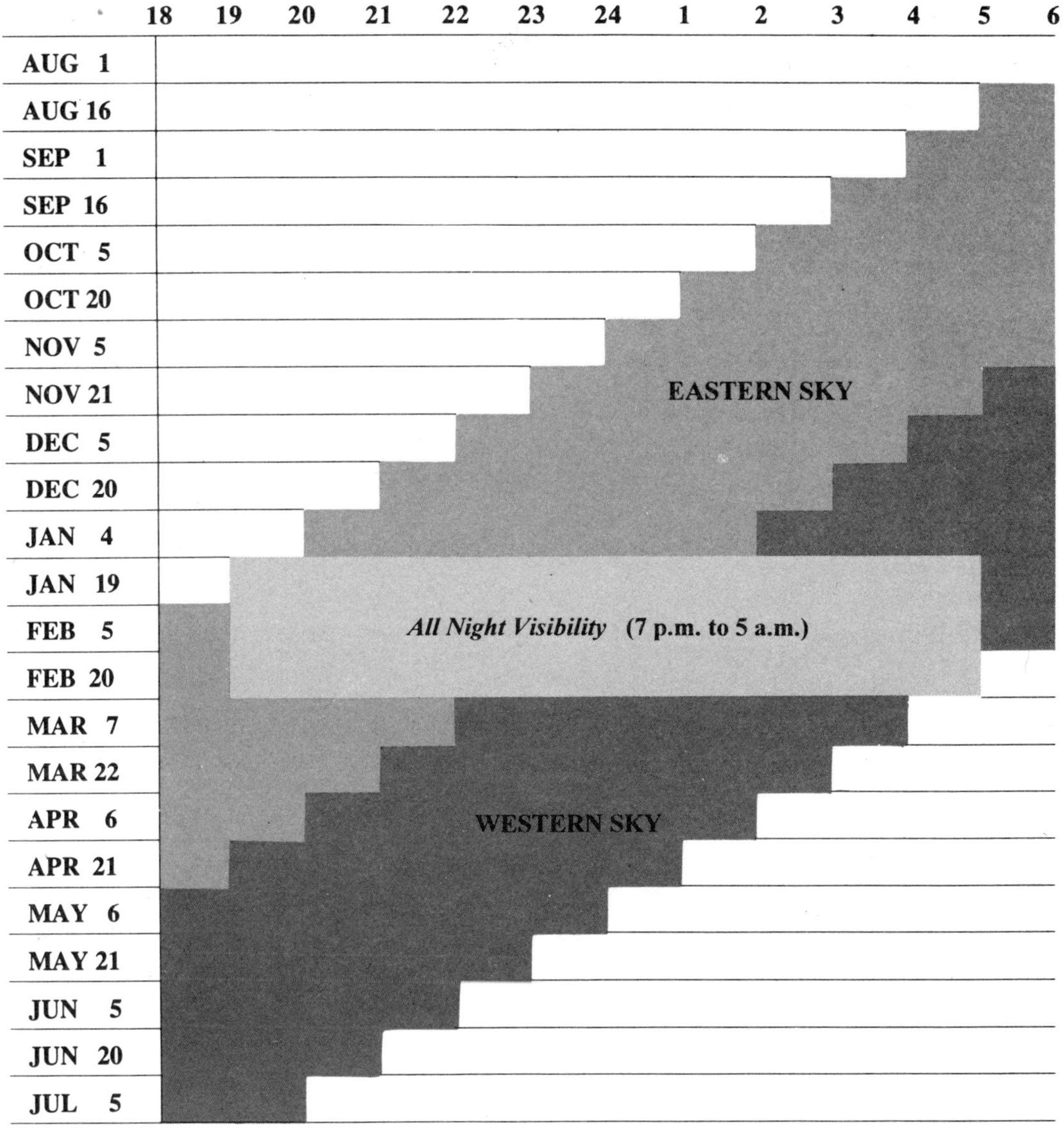

18 19 20 21 22 23 24 1 2 3 4 5 6
AUG 1
AUG 16
SEP 1
SEP 16
OCT 5
OCT 20
NOV 5
NOV 21
DEC 5
DEC 20
JAN 4
JAN 19
FEB 5
FEB 20
MAR 7
MAR 22
APR 6
APR 21
MAY 6
MAY 21
JUN 5
JUN 20
JUL 5
EASTERN SKY
All Night Visibility (7 p.m. to 5 a.m.)
WESTERN SKY

MAGHA

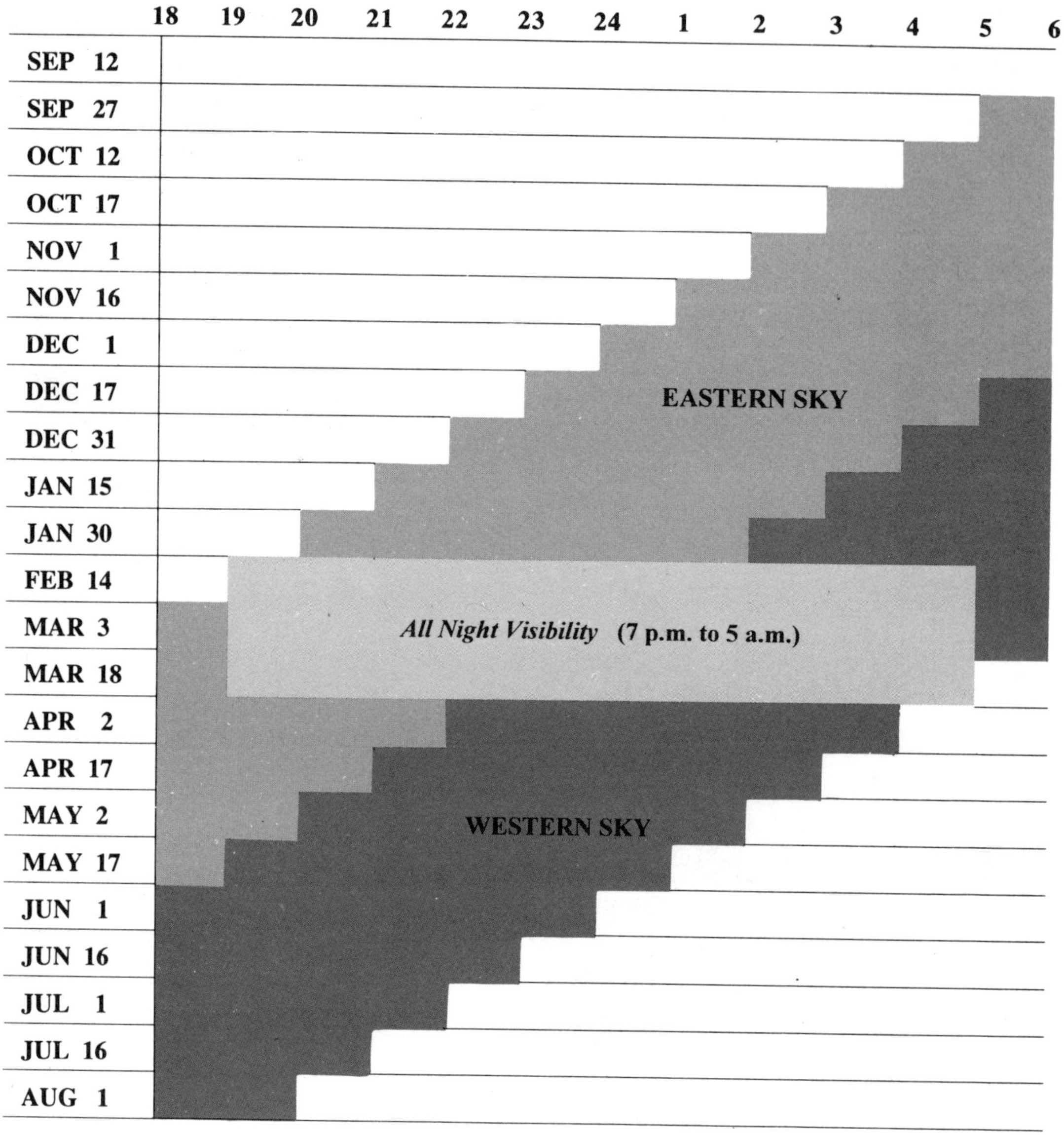
18
19
20
21
22
23
24
1
2
3
4
5
6
SEP 12
SEP 27
OCT 12
OCT 17
NOV 1
NOV 16
DEC 1
DEC 17
DEC 31
JAN 15
JAN 30
FEB 14
MAR 3
MAR 18
APR 2
APR 17
MAY 2
MAY 17
JUN 1
JUN 16
JUL 1
JUL 16
AUG 1
EASTERN SKY
All Night Visibility (7 p.m. to 5 a.m.)
WESTERN SKY

PURVA PHALGUNI

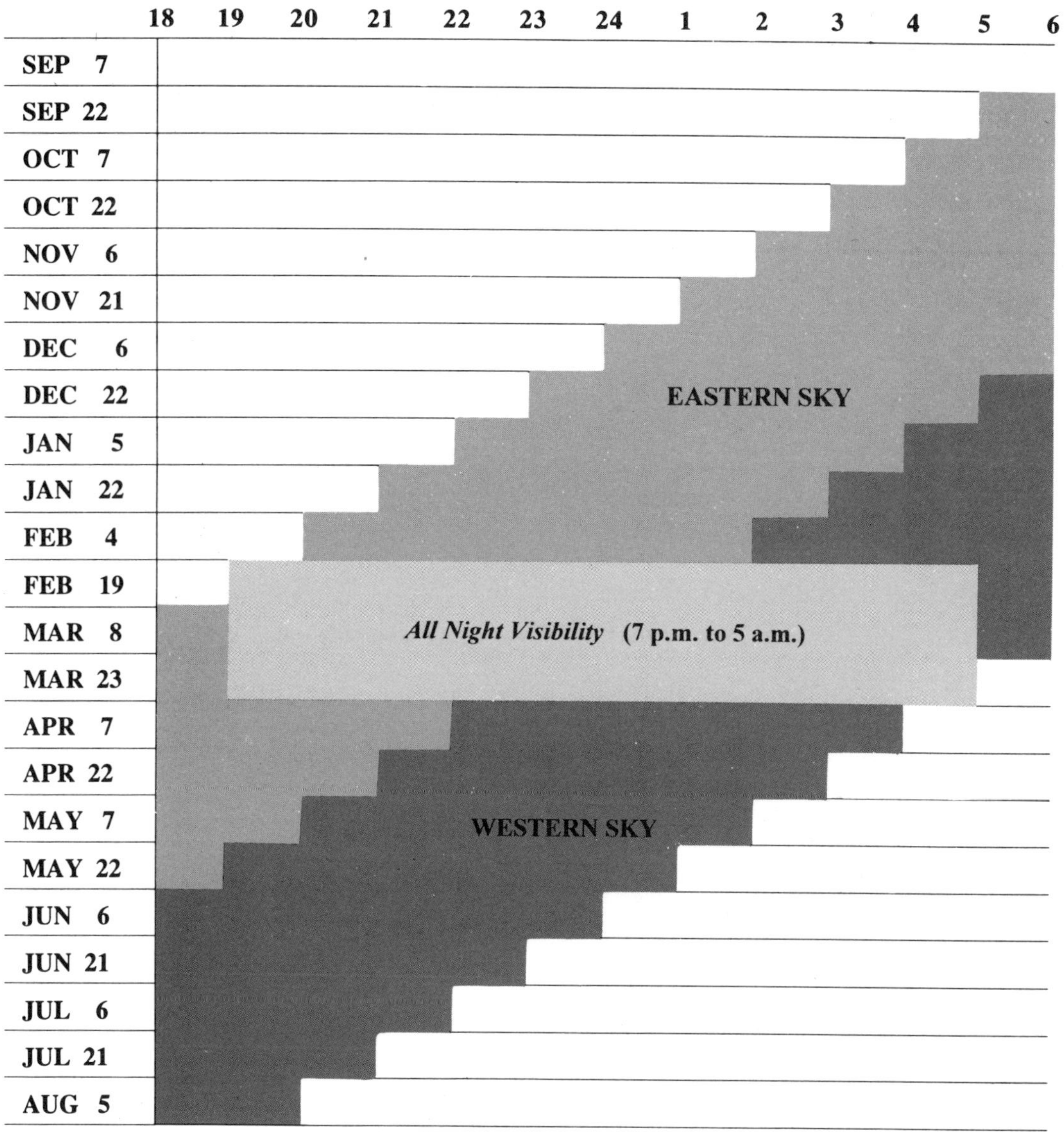

UTTARA PHALGUNI

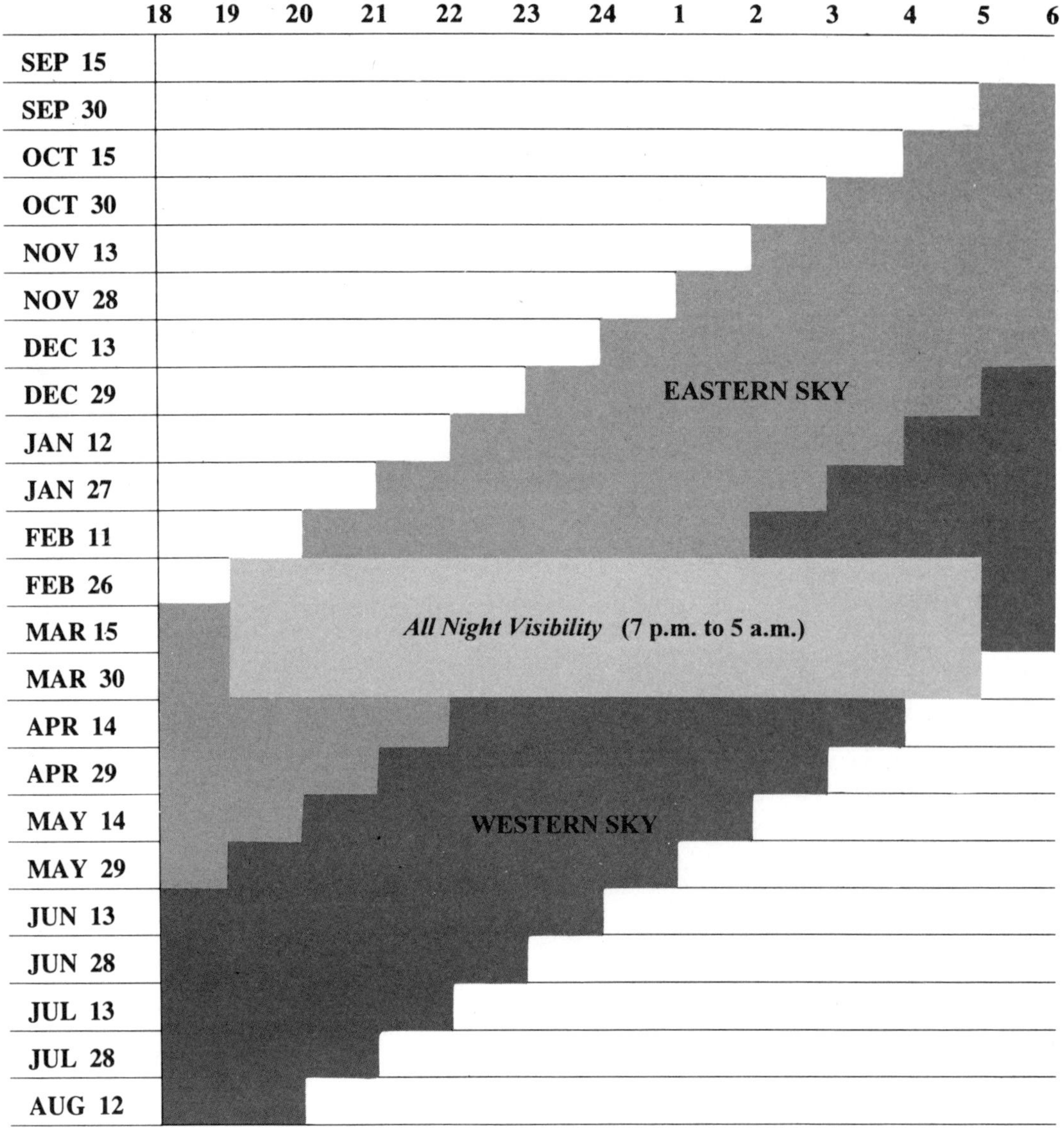

HASTAM

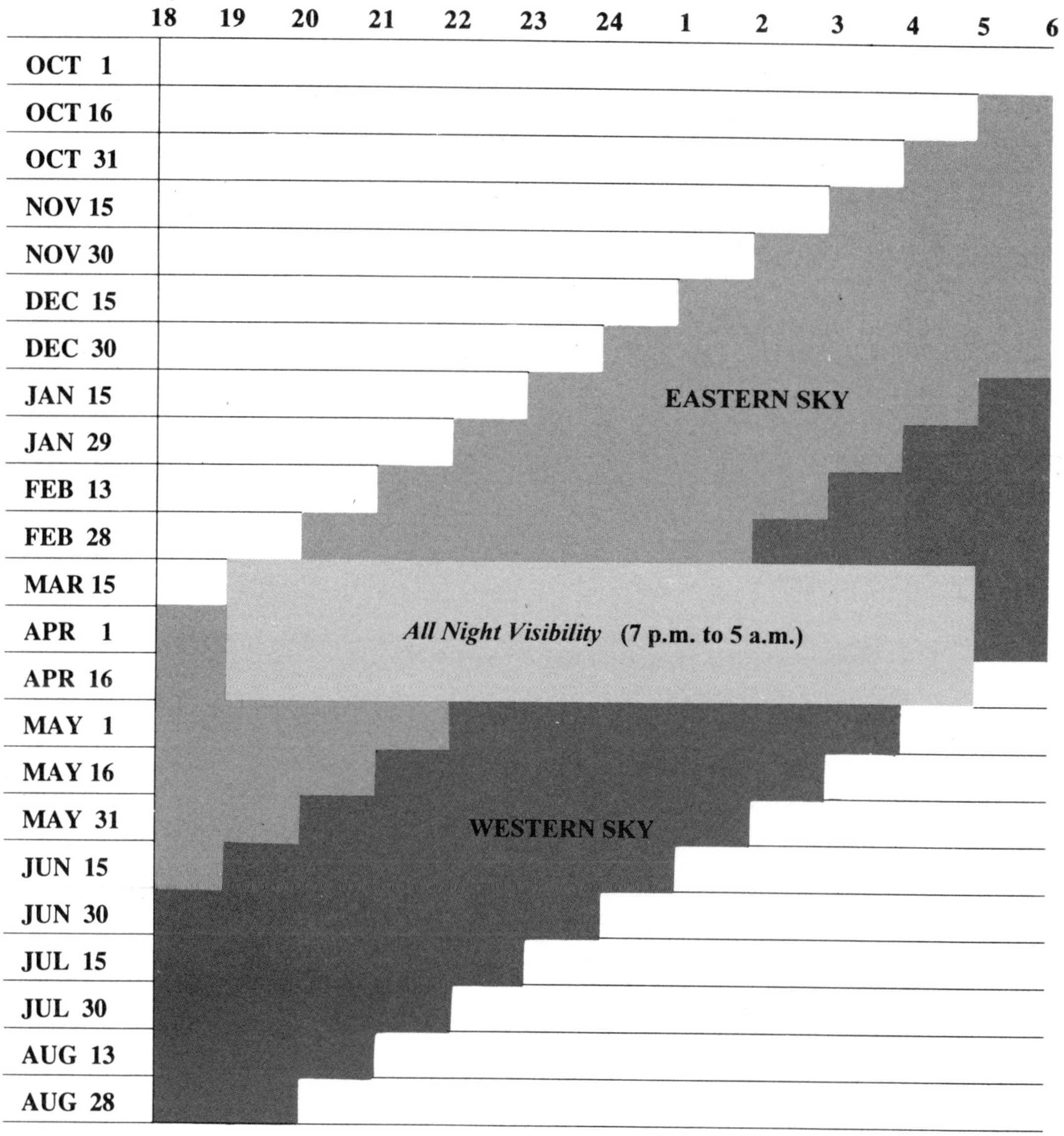
18 19 20 21 22 23 24 1 2 3 4 5 6
OCT 1
OCT 16
OCT 31
NOV 15
NOV 30
DEC 15
DEC 30
JAN 15
JAN 29
FEB 13
FEB 28
MAR 15
APR 1
APR 16
MAY 1
MAY 16
MAY 31
JUN 15
JUN 30
JUL 15
JUL 30
AUG 13
AUG 28
EASTERN SKY
All Night Visibility (7 p.m. to 5 a.m.)
WESTERN SKY

CHITRA

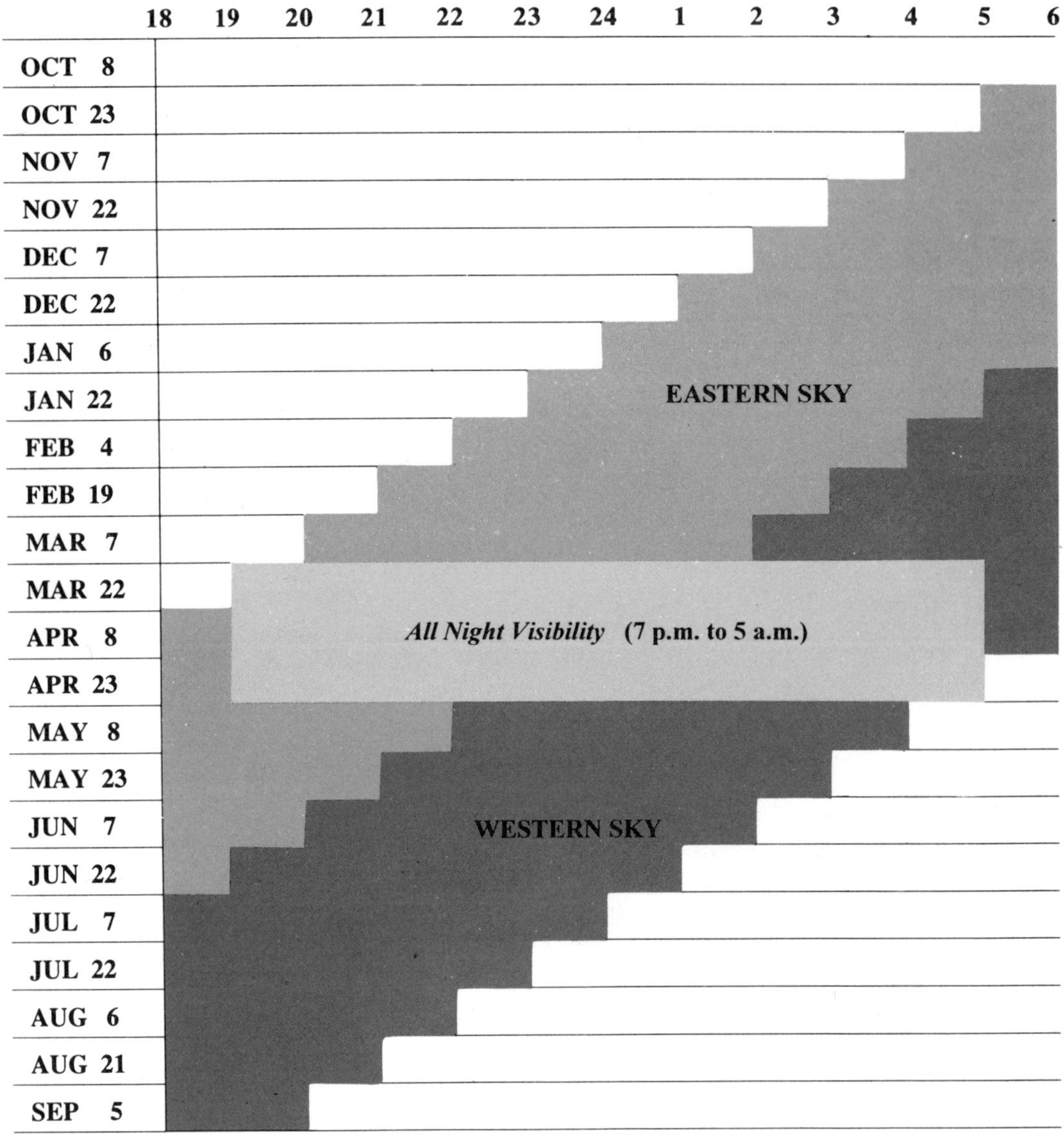

18 19 20 21 22 23 24 1 2 3 4 5 6
OCT 8
OCT 23
NOV 7
NOV 22
DEC 7
DEC 22
JAN 6
JAN 22
FEB 4
FEB 19
MAR 7
MAR 22
APR 8
APR 23
MAY 8
MAY 23
JUN 7
JUN 22
JUL 7
JUL 22
AUG 6
AUG 21
SEP 5
EASTERN SKY
All Night Visibility (7 p.m. to 5 a.m.)
WESTERN SKY

SWATI

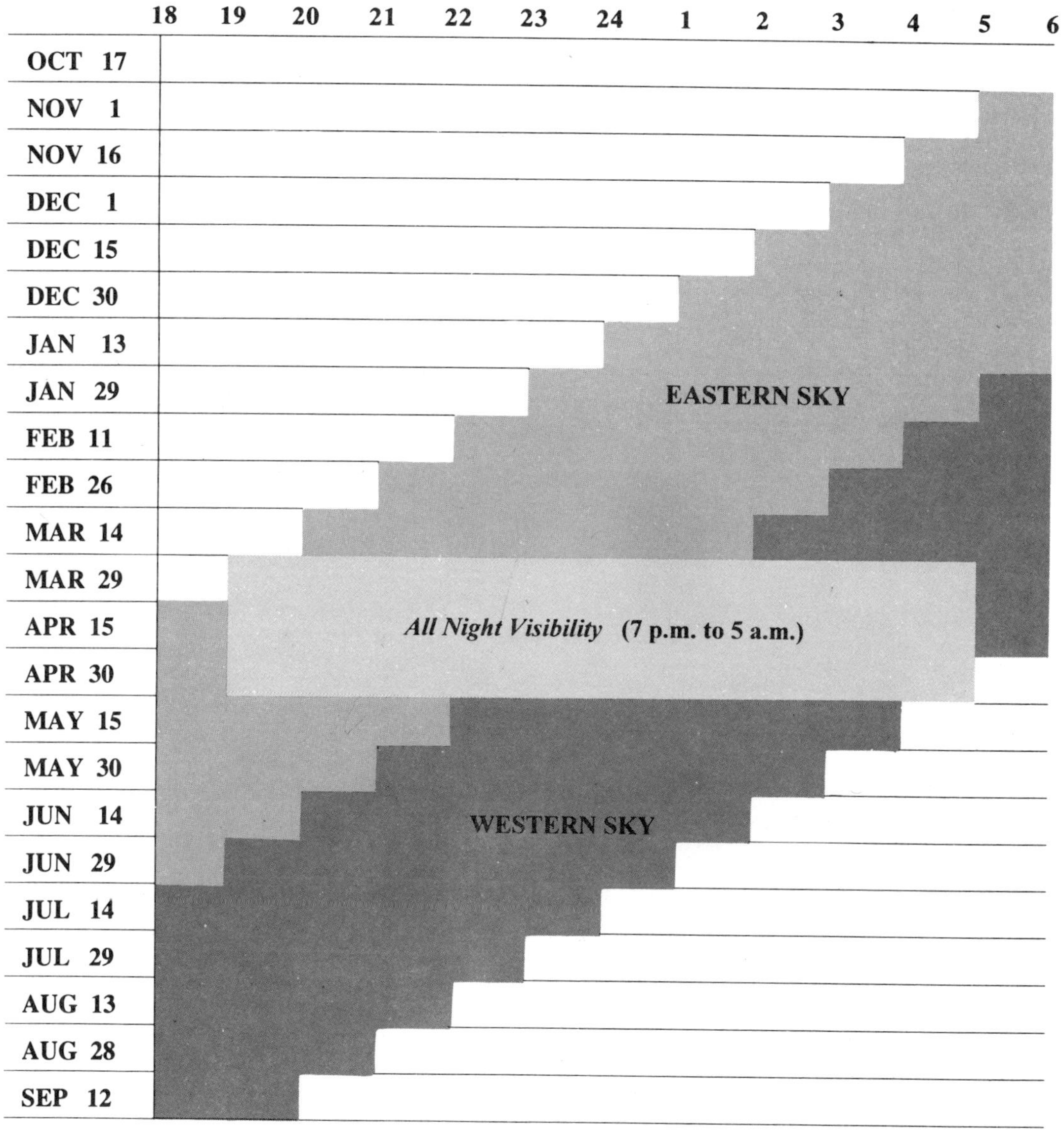
18
19
20
21
22
23
24
1
2
3
4
5
6
OCT 17
NOV 1
NOV 16
DEC 1
DEC 15
DEC 30
JAN 13
JAN 29
FEB 11
FEB 26
MAR 14
MAR 29
APR 15
APR 30
MAY 15
MAY 30
JUN 14
JUN 29
JUL 14
JUL 29
AUG 13
AUG 28
SEP 12
EASTERN SKY
All Night Visibility (7 p.m. to 5 a.m.)
WESTERN SKY

VIS AKA

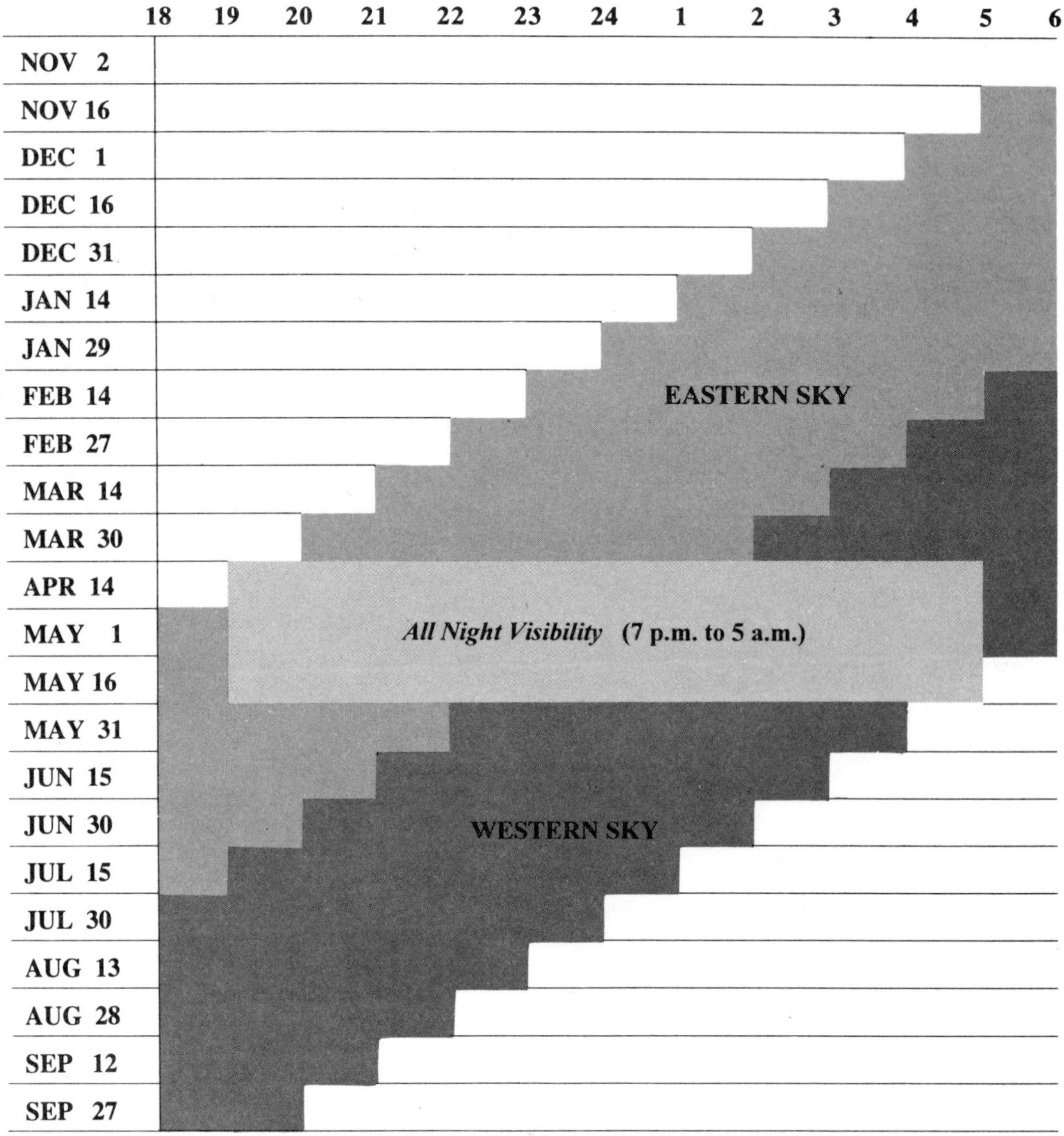
18
19
20
21
22
23
24
1
2
3
4
5
6
NOV 2
NOV 16
DEC 1
DEC 16
DEC 31
JAN 14
JAN 29
FEB 14
FEB 27
MAR 14
MAR 30
APR 14
MAY 1
MAY 16
MAY 31
JUN 15
JUN 30
JUL 15
JUL 30
AUG 13
AUG 28
SEP 12
SEP 27
EASTERN SKY
All Night Visibility (7 p.m. to 5 a.m.)
WESTERN SKY

ANURADHA

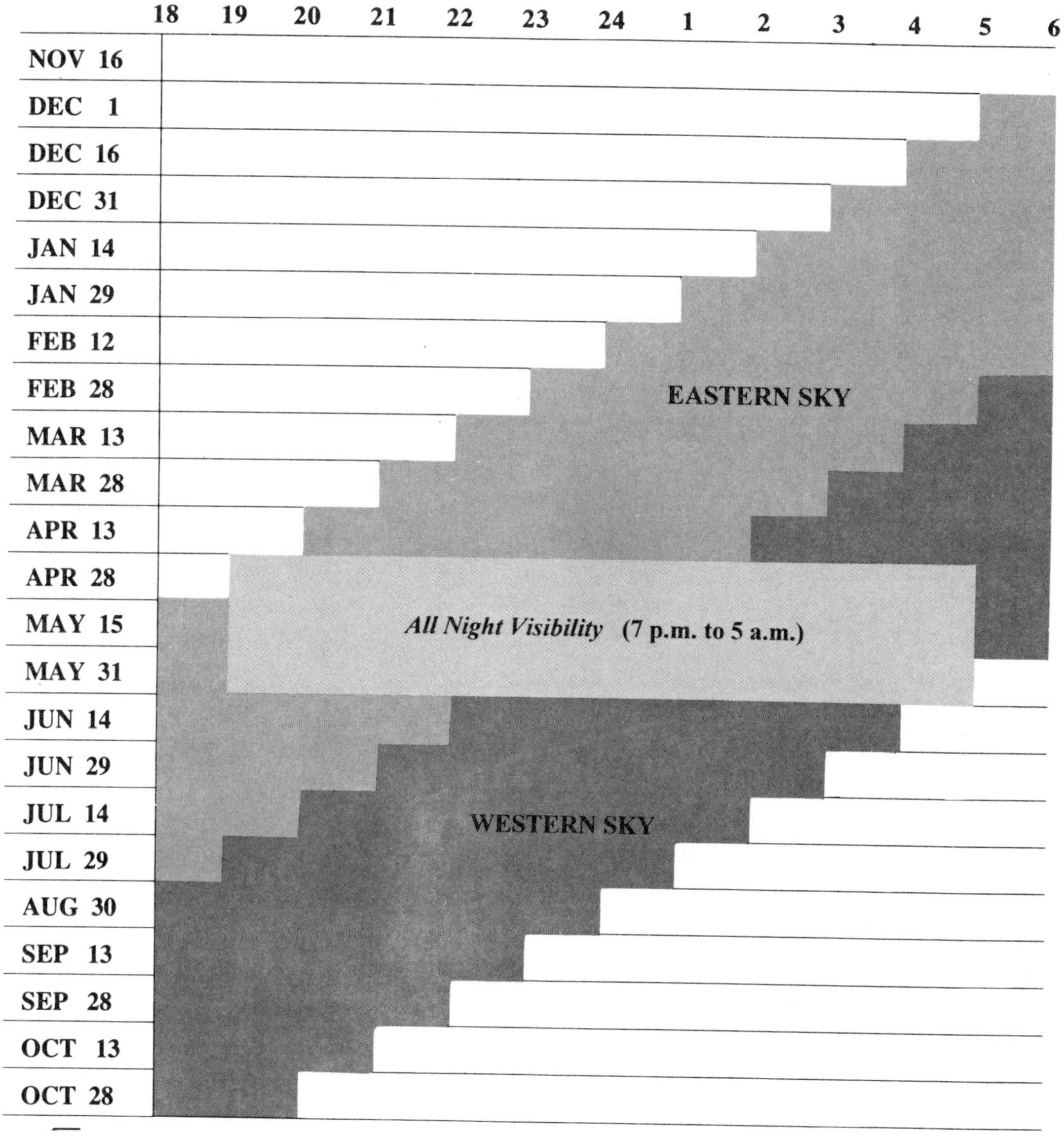
18
19
20
21
22
23
24
1
2
3
4
5
6
NOV 16
DEC 1
DEC 16
DEC 31
JAN 14
JAN 29
FEB 12
FEB 28
MAR 13
MAR 28
APR 13
APR 28
MAY 15
MAY 31
JUN 14
JUN 29
JUL 14
JUL 29
AUG 30
SEP 13
SEP 28
OCT 13
OCT 28
EASTERN SKY
All Night Visibility (7 p.m. to 5 a.m.)
WESTERN SKY

JYESHTA

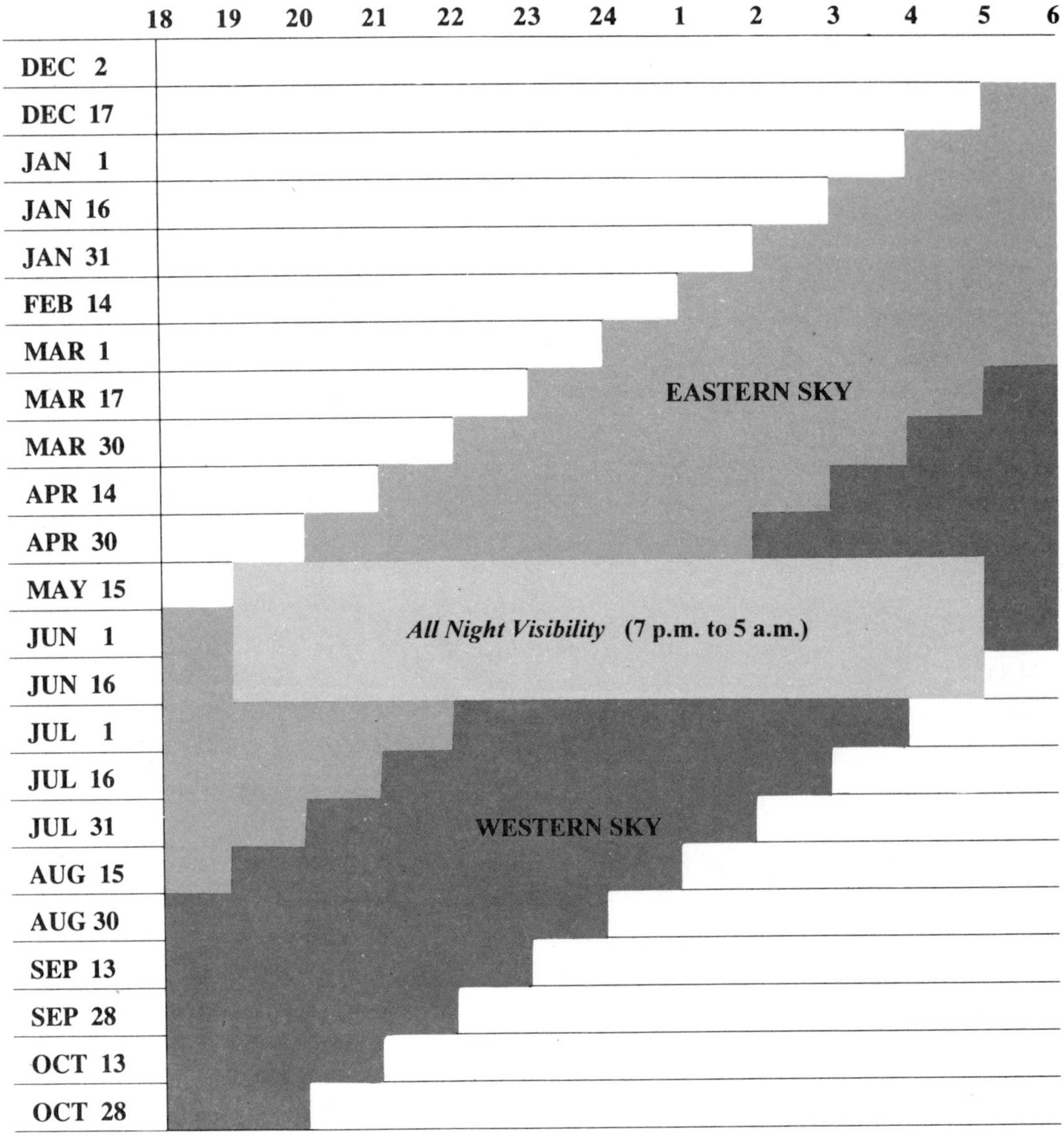

18
19
20
21
22
23
24
1
2
3
4
5
6
DEC 2
DEC 17
JAN 1
JAN 16
JAN 31
FEB 14
MAR 1
MAR 17
MAR 30
APR 14
APR 30
MAY 15
JUN 1
JUN 16
JUL 1
JUL 16
JUL 31
AUG 15
AUG 30
SEP 13
SEP 28
OCT 13
OCT 28
EASTERN SKY
All Night Visibility (7 p.m. to 5 a.m.)
WESTERN SKY

MOOLAM

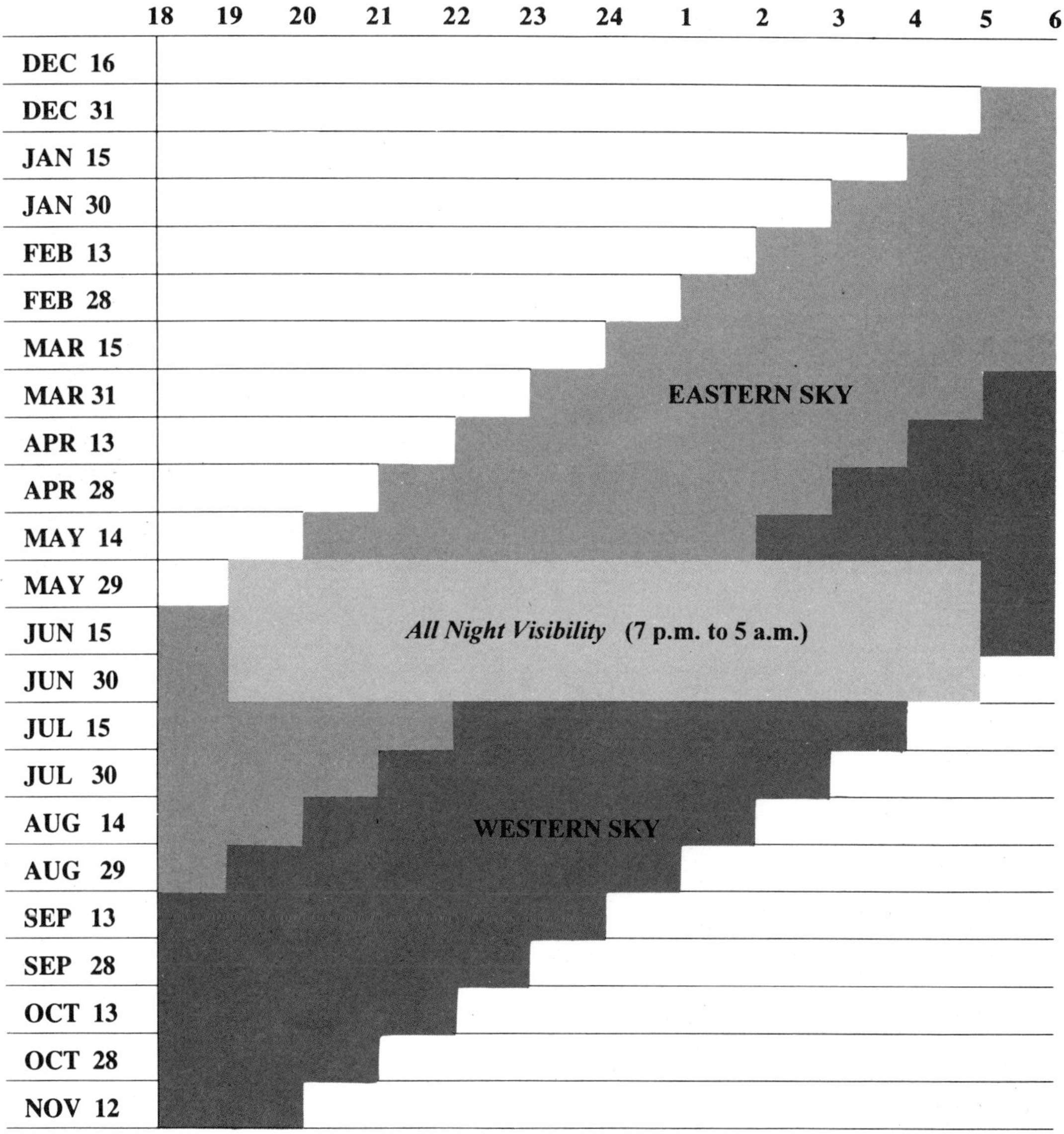
18
19
20
21
22
23
24
1
2
3
4
5
6
DEC 16
DEC 31
JAN 15
JAN 30
FEB 13
FEB 28
MAR 15
MAR 31
APR 13
APR 28
MAY 14
MAY 29
JUN 15
JUN 30
JUL 15
JUL 30
AUG 14
AUG 29
SEP 13
SEP 28
OCT 13
OCT 28
NOV 12
EASTERN SKY
All Night Visibility (7 p.m. to 5 a.m.)
WESTERN SKY

PURVASHADHA

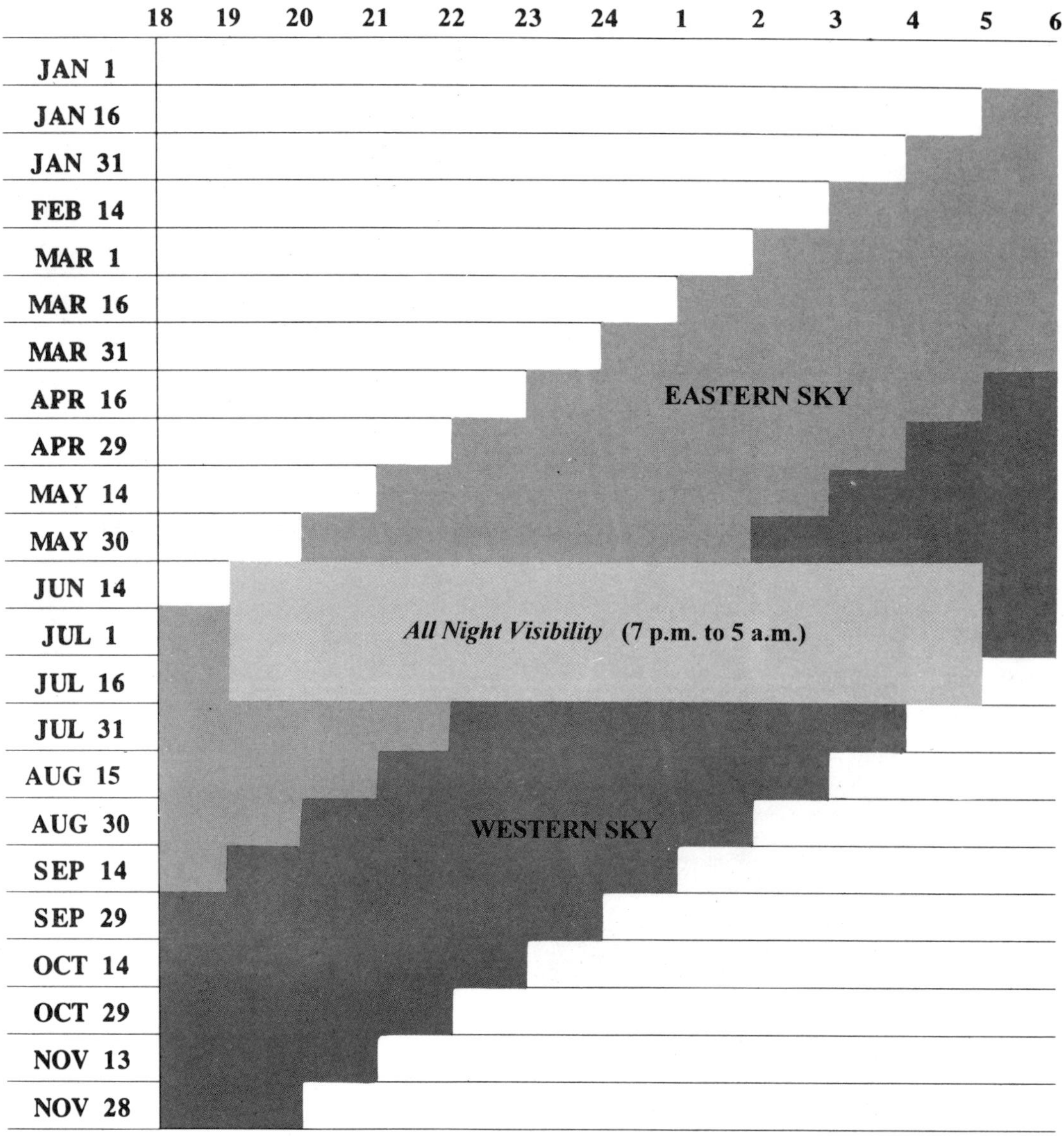

UTTARASHADHA

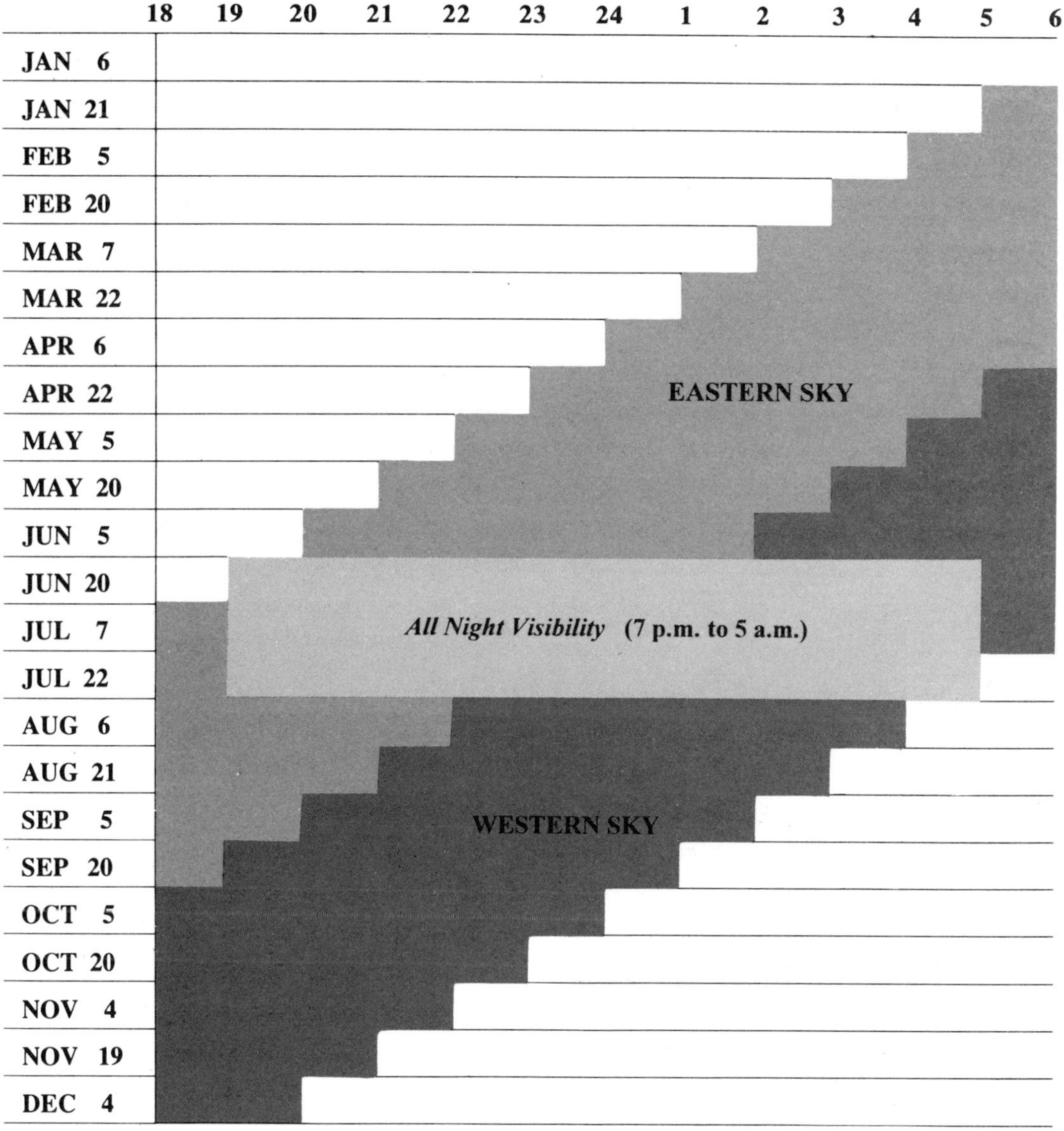

SRAVANA

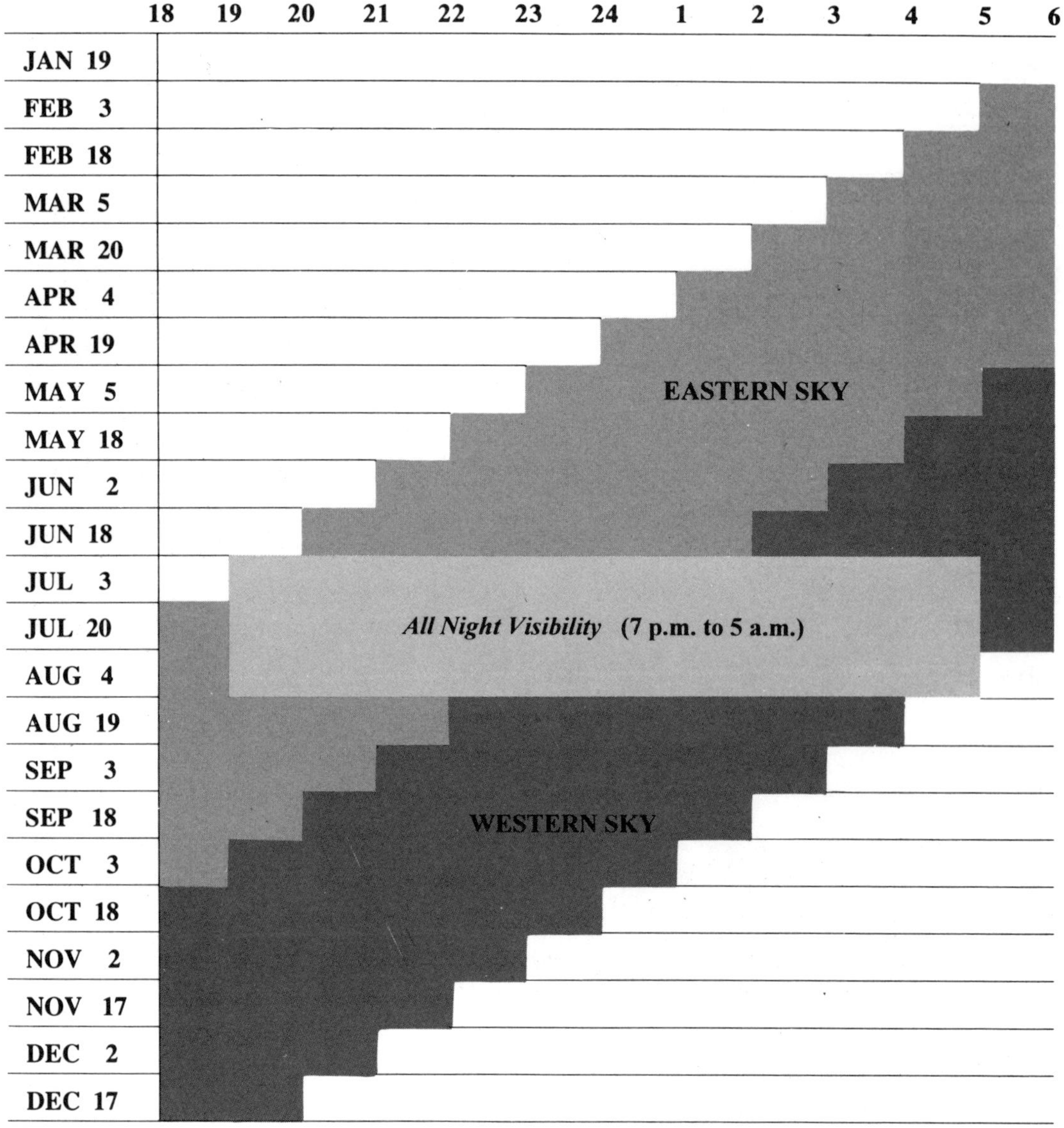
18
19
20
21
22
23
24
1
2
3
4
5
6
JAN 19
FEB 3
FEB 18
MAR 5
MAR 20
APR 4
APR 19
MAY 5
MAY 18
JUN 2
JUN 18
JUL 3
JUL 20
AUG 4
AUG 19
SEP 3
SEP 18
OCT 3
OCT 18
NOV 2
NOV 17
DEC 2
DEC 17
EASTERN SKY
All Night Visibility (7 p.m. to 5 a.m.)
WESTERN SKY

DHANISHTA

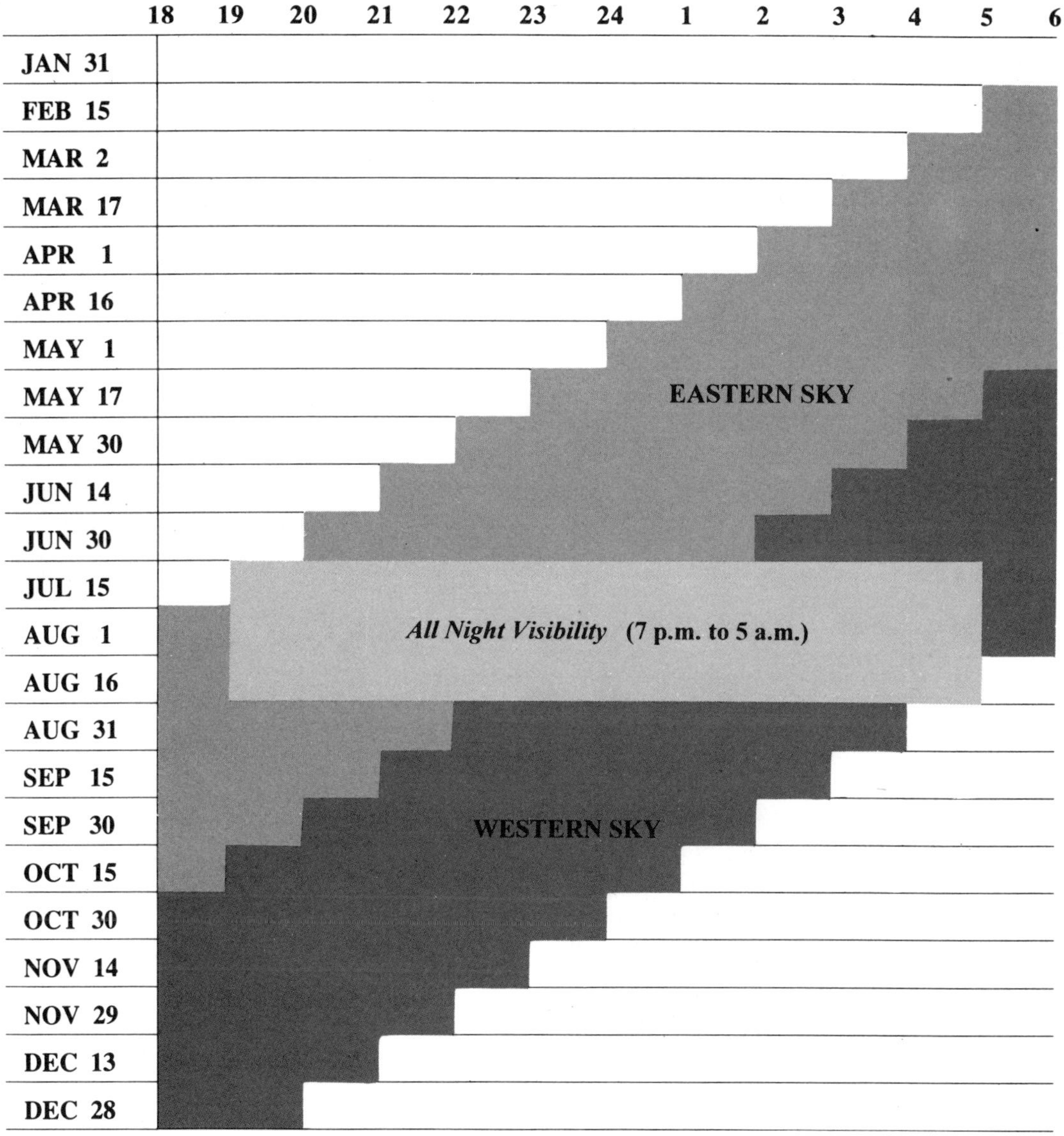

18
19
20
21
22
23
24
1
2
3
4
5
6
JAN 31
FEB 15
MAR 2
MAR 17
APR 1
APR 16
MAY 1
MAY 17
MAY 30
JUN 14
JUN 30
JUL 15
AUG 1
AUG 16
AUG 31
SEP 15
SEP 30
OCT 15
OCT 30
NOV 14
NOV 29
DEC 13
DEC 28
EASTERN SKY
All Night Visibility (7 p.m. to 5 a.m.)
WESTERN SKY

SATABHISHAJ

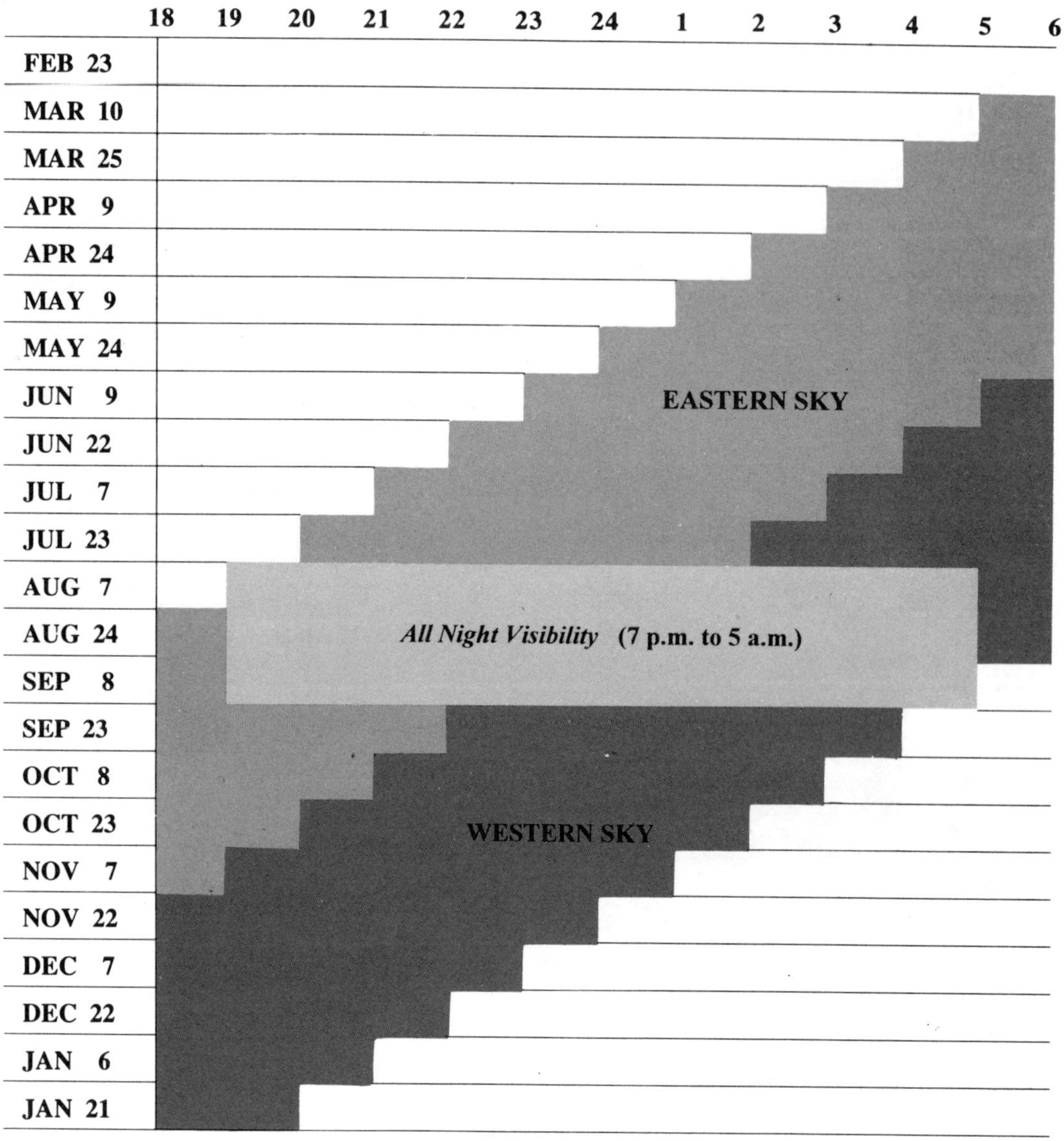

PURVA-PROSHTAPADA

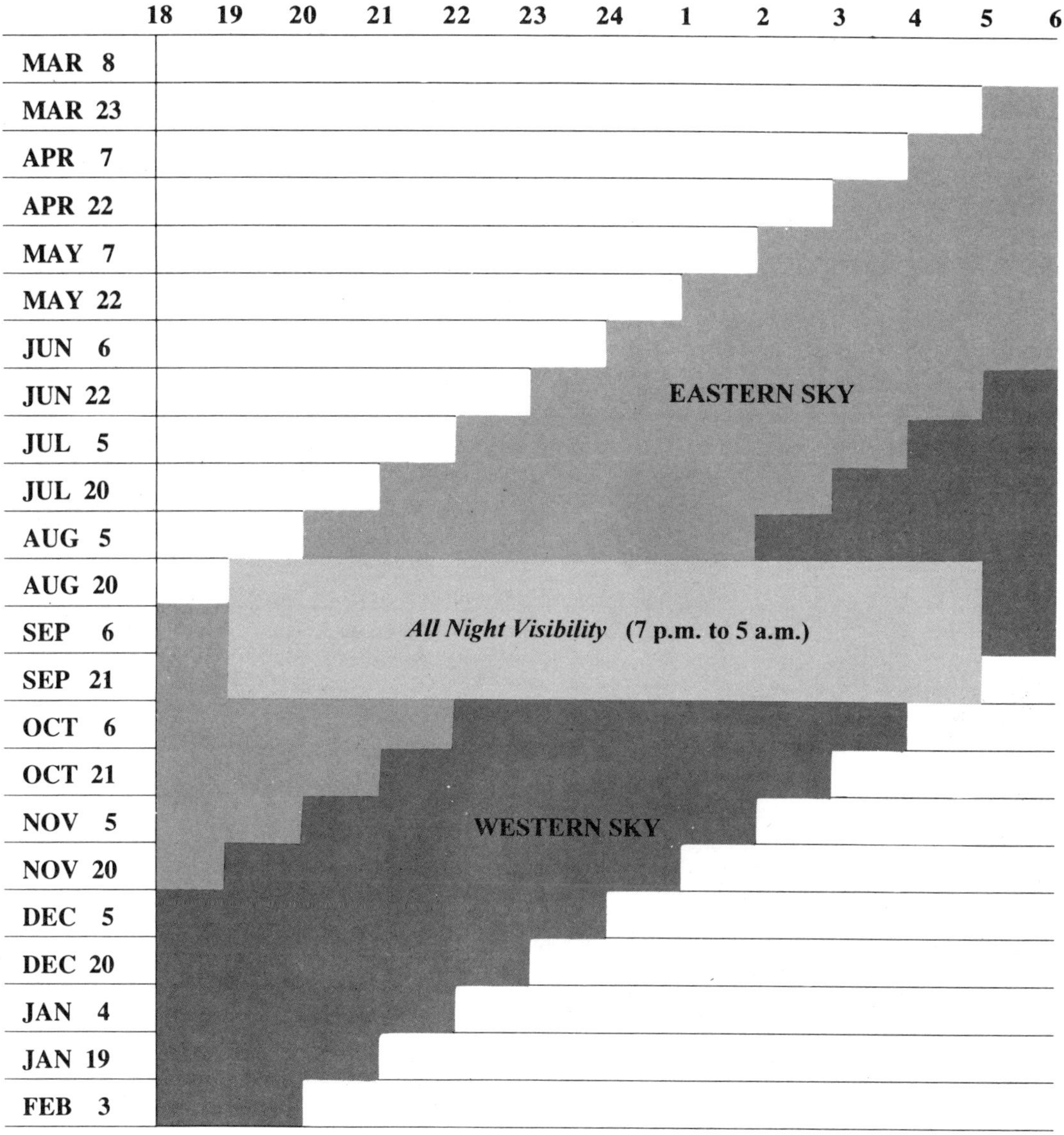

UTTARA-PROSHTAPADA

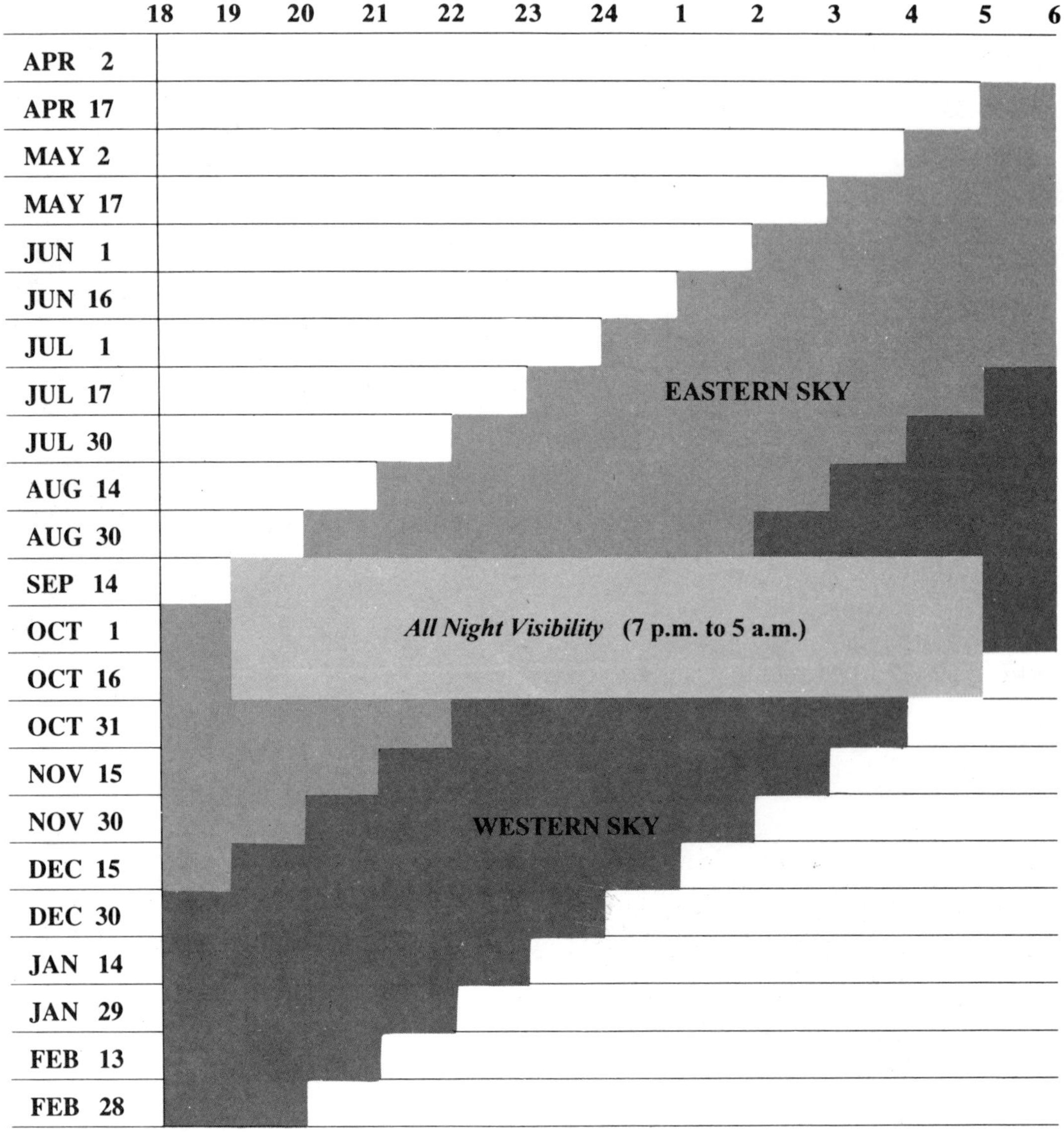

REVATI

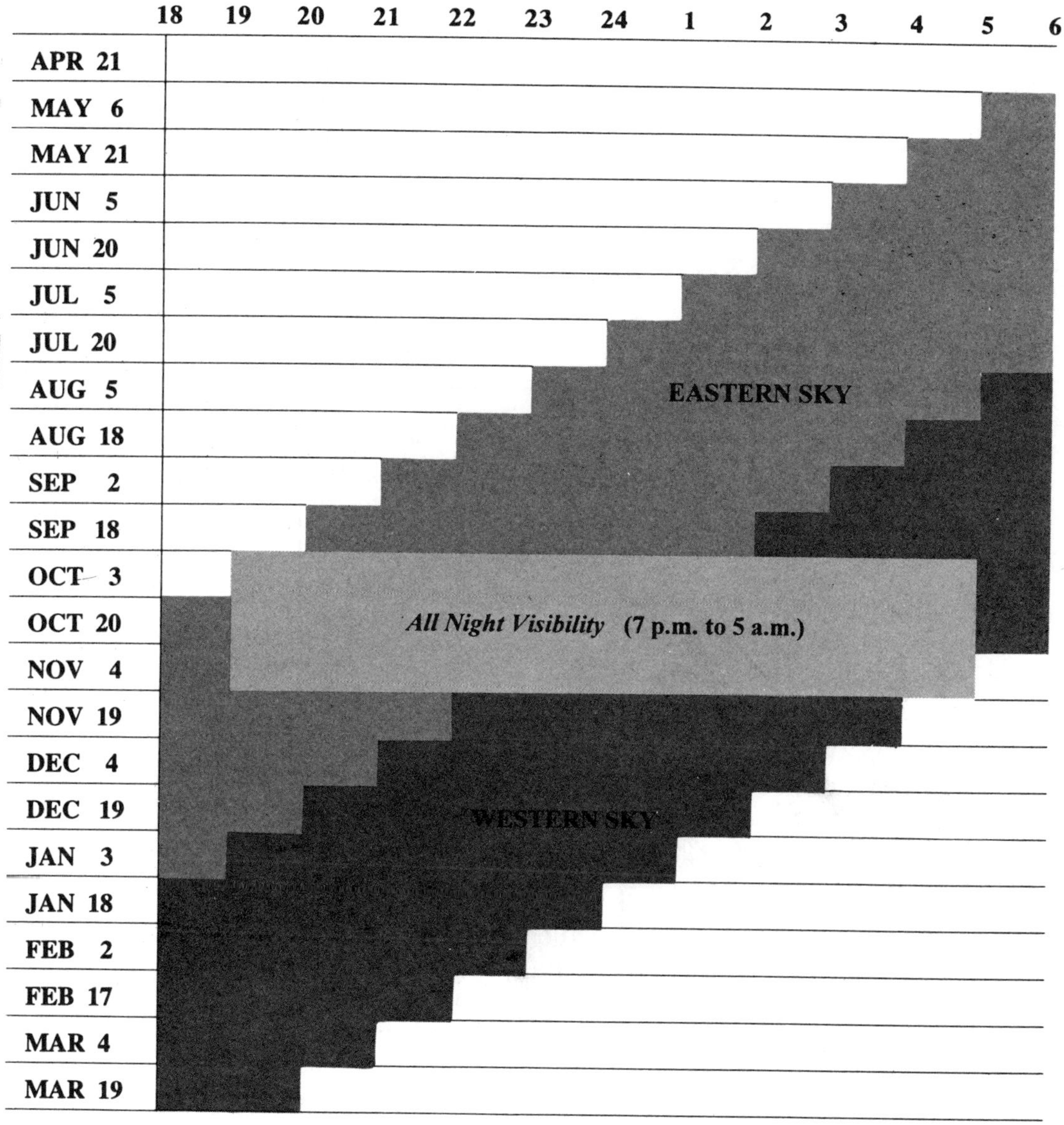

18
19
20
21
22
23
24
1
2
3
4
5
6
APR 21
MAY 6
MAY 21
JUN 5
JUN 20
JUL 5
JUL 20
AUG 5
AUG 18
SEP 2
SEP 18
OCT 3
OCT 20
NOV 4
NOV 19
DEC 4
DEC 19
JAN 3
JAN 18
FEB 2
FEB 17
MAR 4
MAR 19
EASTERN SKY
All Night Visibility (7 p.m. to 5 a.m.)
WESTERN SKY

Appendix III

(Ref. page 28)

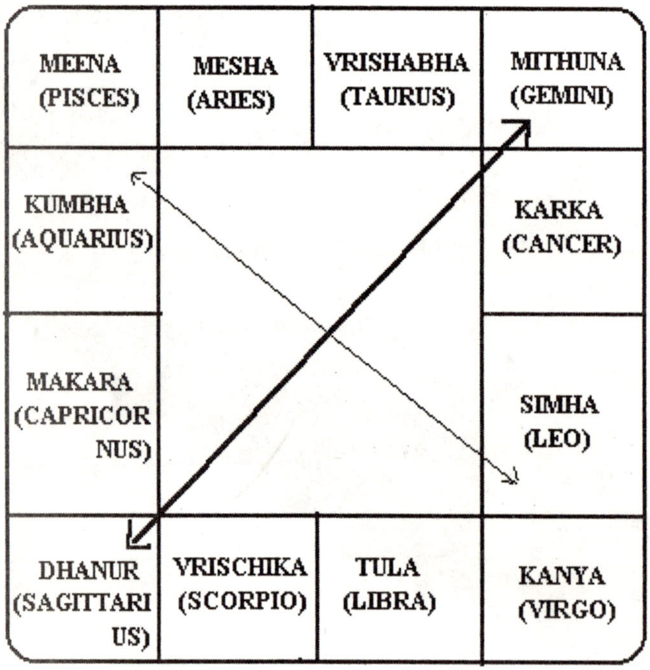

The above chart helps the visualization of the seventh sign from any Zodiacal sign. It is always the diagonally opposite one in the chart. For instance *Mithuna* and *Dhanur* are seventh from each other. *Kumbha* and *Simha* are seventh from each other. So also are *Meena* and *Kanya*, though the diagonal line is not shown. Similarly we have for the other pairs. Incidentally the above chart is the standard way in which the astrologers of the Southern part of India fix the positions of the signs in their representative horoscope charts.

Appendix IV

(Ref. page 38)

MTF FOR THE REMAINING NINE MONTHS

CHITTIRAI MONTH

STAR NAMES	COUNT FROM PUSHYA (N)	NO.OF GHATIKAS PAST SUNSET [N X (9/4)] - 5
PUSHYA	1	-2.75
ÂSLESHA	2	-0.5
MAGHA	3	1.75
P.PHALGUNI	4	4
U.PHALGUNI	5	6.25
HASTÂ	6	8.5
CHITRÂ	7	10.75
SWÂTI	8	13
VISHAKA	9	15.25
ANURADHA	10	17.5
JYESHTÂ	11	19.75
MÛLA	12	22
PURVASHADA	13	24.25
UTTARASHADA	14	26.5
ŠRONA	15	28.75
ŠRAVISHTA	16	31

ÂNI MONTH

STAR NAMES	COUNT FROM PUSHYA (N) [N X (9/4)] - 15	NO.OF GHATIKAS PAST SUNSET
PUSHYA	1	-12.75
ÂSHLESHA	2	-10.5
MAGHA	3	-8.25
P.PHALGUNI	4	-6
U.PHALGUNI	5	-3.75
HASTÂ	6	-1.5
CHITRÂ	7	0.75
SWÂTI	8	3
VISHAKA	9	5.25
ANURADHA	10	7.5
JYESHTÂ	11	9.75
MÛLA	12	12
PURVASHÂDA	13	14.25
UTTARASHÂDA	14	16.5
ŠRONA	15	18.75
ŠRAVISHTA	16	21
ŠATABISHAJ	17	23.25
P.PROSHTAPADA	18	25.5
U.PROSHTAPADA	19	27.75
REVATI	20	30

ÂDI MONTH

STAR NOS.	COUNT FROM PUSHYA (N)	NO.OF GHATIKAS PAST SUNSET [N X (9/4)] - 20
PUSHYA	1	-17.75
ÂSLESHA	2	-15.5
MAGHA	3	-13.25
P/PHALGUNI	4	-11
U.PHALGUNI	5	-8.75
HASTÂ	6	-6.5
CHITRÂ	7	-4.25
SWÂTI	8	-2
VISHAKA	9	0.25
ANURADHA	10	2.5
JYESHATÂ	11	4.75
MÛLA	12	7
PURVASHÂDA	13	9.25
UTTARASHÂDA	14	11.5
ŚRONA	15	13.75
ŚRAVISHTA	16	16
SATABHISHAJ	17	18.25
P.PROSHTAPADA	18	20.5
U.PROSHTAPADA	19	22.75
REVATI	20	25
ASWINI	21	27.25
BHARANI	22	29.5

ÂVANI MONTH

STAR NAMES	COUNT FROM ANURADHA (N)	NO.OF GHATIKAS PAST SUNSET [N x (9/4)] - 5
ANURADHA	1	-2.75
JYESHTÂ	2	-0.5
MÛLA	3	1.75
PURVASHÂDA	4	4
UTTARASHÂDA	5	6.25
ŠRONA	6	8.5
ŠRAVISHTA	7	10.75
SATABHISHAJ	8	13
P.PROSHTAPADA	9	15.25
U.PROSHTAPADA	10	17.5
REVATI	11	19.75
ASWINI	12	22
BHARANI	13	24.25
KRITTIKÂ	14	26.5
ROHINI	15	28.75

PURATTÂSI MONTH

STAR NAMES	COUNT FROM ANURADHA (N)	NO.OF GHATIKAS PAST SUNSET [N x (9/4)] - 10
ANURADHA	1	-7.75
JYESHTÂ	2	-5.5
MÛLA	3	-3.25
PURVASHÂDA	4	-1
UTTARASHÂDA	5	1.25
ŠRONA	6	3.5
ŠRAVISHTA	7	5.75
SATABHISHAJ	8	8
P.PROSHTAPADA	9	10.25
U.PROSHTAPADA	10	12.5
REVATI	11	14.75
ASWINI	12	17
BHARANI	13	19.25
KRITTIKÂ	14	21.5
ROHINI	15	23.75
MRIGAŠIRÂH	16	26
ÂRDRÂ	17	28.25
PUNARVASU	18	30.5

KÂRTIGAI MONTH

STAR NAMES	COUNT FROM ANURADHA (N)	NO.OF GHATIKAS PAST SUNSET [N x (9/4)] - 20
ANURADHA	1	-17.75
JYESHTÂ	2	-15.5
MÛLA	3	-13.25
PURVASHÂDA	4	-11
UTTARASHÂDA	5	-8.75
ŠRONA	6	-6.5
ŠRAVISHTA	7	-4.25
SATABHISHAJ	8	-2
P.PROSHTAPADA	9	0.25
U.PROSHTAPADA	10	2.5
REVATI	11	4.75
ASWINI	12	7
BHARANI	13	9.25
KRITTIKÂ	14	11.5
ROHINI	15	13.75
MRIGAŠIRÂH	16	16
ÂRDRÂ	17	18.25
PUNARVASU	18	20.5
ÂSLESHA	19	22.75
MAGHA	20	25
P.PHALGUNI	21	27.25
U.PHALGUNI	22	29.5

MÂRGAZHI MONTH

STAR NAMES	COUNT FROM U.PROSHTAPADA (N)	NO.OF GHATIKAS PAST SUNSET [N x (9/4)] - 5
U.PROSHTAPADA	1	-2.75
REVATI	2	-0.5
ASWINI	3	1.75
BHARANI	4	4
KRITTIKÂ	5	6.25
ROHINI	6	8.5
MRIGAŠIRÂH	7	10.75
ÂRDRÂ	8	13
PUNARVASU	9	15.25
PUSHYA	10	17.5
ÂSHLESHA	11	19.75
MAGHA	12	22
P.PHALGUNI	13	24.25
U.PHALGUNI	14	26.5
HASTÂ	15	28.75

TAI MONTH

STAR NAMES	COUNT FROM U.PROSHTAPADA (N)	NO.OF GHATIKAS PAST SUNSET [N x (9/4)] - 10
U.PROSHTAPADA	1	-7.75
REVATI	2	-5.5
ASWINI	3	-3.25
BHARANI	4	-1
KRITTIKÂ	5	1.25
ROHINI	6	3.5
MRIGAŚIRÂH	7	5.75
ÂRDRÂ	8	8
PUNARVASU	9	10.25
PUSHYA	10	12.5
ÂSHLESHA	11	14.75
MAGHA	12	17
P.PHALGUNI	13	19.25
U.PHALGUNI	14	21.5
HASTÂ	15	23.75
CHITRÂ	16	26
SWÂTI	17	28.25
VISHAKA	18	30.5

MÂSI MONTH

STAR NAMES	COUNT FROM U. PROSHTAPADA (N)	NO.OF GHATIKAS PAST SUNSET [N x (9/4)] - 15
U.PROSHTAPADA	1	-12.75
REVATI	2	-10.5
ASWINI	3	-8.25
BHARANI	4	-6
KRITTIKÂ	5	-3.75
ROHINI	6	-1.5
MRIGAŚIRÂH	7	0.75
ÂRDRÂ	8	3
PUNARVASU	9	5.25
PUSHYA	10	7.5
ÂSHLESHA	11	9.75
MAGHA	12	12
P.PHALGUNI	13	14.25
U.PHALGUNI	14	16.5
HASTÂ	15	18.75
CHITRÂ	16	21
SWÂTI	17	23.25
VISHAKA	18	25.5
ANURADHA	19	27.75
JYESHTÂ	20	30

Appendix V

CONCORDANCE BETWEEN NAMES OF STARS IN SANSKRIT, TAMIL AND ENGLISH

Sanskrit	Tamil	English
Ašvinî	Achvinî	Hamal
Bharanî	Bharanî	τ – Arietis
Krittikâ	Kârtigai	Pleides
Rohinî	Rohinî	Aldebaran
Mrigaširâh	Mirugasirsham = Mânralai	λ - Orionis
Ârdrâ	Tiruvadirai	Betelgeuse
Punarvasu	Punarpûsam	Castor & Pollux
Pushya	Pûsam	Praesepe cluster
Âsleshâ	Âyilyam	ε, ρ, ζ, δ – Hydra
Maghâ	Magam	Regulus
Pûrva-phalguni	Pûram	Chertan
Uttara-phalguni	Uttiram	Denebola
Hastâ	Hastam	δ, γ, β, ε, α – Corvus
Chitrâ	Chittirai	Spica
Swâti	Suvati	Arcturus
Višakâ	Visakam	(Six stars in Libra) See description on page 19
Anurâdha	Anusham	β – Scorpii
Jyeshtâ	Kettai	Antares
Mûla	Mûlam	λ - Scorpii = Shaula

Pûrvâshâdha	Pûradam	This and the next asterism form Four Stars in Sagittarius (See description on page 19)
Uttarâshâdha	Uttirâdam	
Šrona = Šravana	Tiruvonam	Altair
Šravishtâ = Dhanishtâ	Avittam	α – Delphinus
Šatabhishaj	Sadayam	π,ζ, η, γ – Aquarium
Pûrva-Proshtapadâ (also) Pûrva-Bhâdrapadâ	Pûrattadi	β - Pegasus
Uttara-Proshtapadâ (also) Pûrva–Bhâdrapadâ	Uttirattadi	α - Pegasus
Revatî	Revati	Five stars in Pisces (See sketch on page 21)

References

K.D. Abhyankar, Astro Physics - Stars and Galaxies , Tata Mcgraw-Hill Publishing Company Limited , New Delhi (1992) .

Bose, Sen and Subbarayappa, A concise History of Science in India, Indian National Science Academy, New Delhi (1971).

Robert Sewell and Balakrishna Dikshit, The Indian Calendar, Motilal Banarsidass Publishers Delhi (1995).

Menzel, Pasachoff, A Field Guide to the Stars and Planets, Houghton Mifflin Company, Boston (1983).

V.Krishnamurthy, Culture, Excitement and Relevance of Mathematics, Wiley Eastern Limited, New Delhi (1990).

Rajbali Pandey, Hindu Samskaras — Socio-Religious study of the Hindu Sacraments, Motilal Banarsidass Publishers Private Limited Delhi (1994).

H.A. Rey, The Stars - A new way to see them, Houghton Mifflin Company, Boston (1952).

Tulaja Raja, (Manuscript) *Inakularajatejonidhih*, Burnell's Catalogue No. 4263, Tanjore Saraswati Mahal Library (circa 1728).

Vedavyas, Astronomical dating of the Mahabharata war, Vedavyas Bharati, Sriparvatham, Gowripatnam W.G. Dist, A.P. India, 2nd edition (1995).

F.Zigel, Wonders of the night sky, Mir Publishers, Moscow (1968).

Index

of names or technical terms and their first/major appearances